MOTION UNDERSTANDING
Robot and Human Vision

THE KLUWER INTERNATIONAL SERIES IN
ENGINEERING AND COMPUTER SCIENCE

ROBOTICS: VISION, MANIPULATION AND SENSORS

Consulting Editor

Takeo Kanade
Carnegie Mellon University

Other books in the series:

Robotic Grasping and Fine Manipulation, M. Cutkosky
ISBN 0-89838-200-9

Shadows and Silhouettes in Computer Vision, S. Shafer
ISBN 0-89838-167-3

Perceptual Organization and Visual Recognition, D. Lowe
ISBN 0-89838-172-X

Robot Dynamics Algorithms, R. Featherstone
ISBN 0-89838-230-0

Three Dimensional Machine Vision, T. Kanade, ed.
ISBN 0-89838-188-6

Kinematic Modeling Identification, and Control of Robotic Manipulators,
H.W. Stone
ISBN 0-89838-237-8

Robotic Object Recognition Using Vision and Touch, P. Allen
ISBN 0-89838-245-9

Integration, Coordination and Control of Multi-Sensor Robot Systems,
H.F. Durrant-Whyte
ISBN 0-89838-247-5

MOTION UNDERSTANDING

Robot and Human Vision

edited by

W.N. Martin
University of Virginia

J.K. Aggarwal
University of Texas at Austin

KLUWER ACADEMIC PUBLISHERS
Boston/Dordrecht/Lancaster

Distributors for North America:
Kluwer Academic Publishers
101 Philip Drive
Assinippi Park
Norwell, Massachusetts 02061 USA

Distributors for the UK and Ireland:
Kluwer Academic Publishers
MTP Press Limited
Falcon House, Queen Square
Lancaster LA1 1RN, UNITED KINGDOM

Distributors for all other countries:
Kluwer Academic Publishers Group
Distribution Centre
Post Office Box 322
3300 AH Dordrecht, THE NETHERLANDS

Library of Congress Cataloging-in-Publication Data

Motion understanding : robot and human vision / edited by W.N. Martin,
    J.K. Aggarwal.
        p. cm.—(The Kluwer international series in engineering and
computer science ; SECS44)
    Includes bibliographies and indexes.
    ISBN 0–89838–258–0
    1. Robot vision. 2. Vision. 3. Motion. I. Martin, W.N.
(Worthy N.) II. Aggarwal, J.K. III. Series.
TJ211.3.M67 1988   87–31114
629.8′92—dc19   CIP

Printed in the United States of America

This book is dedicated
to Shanti, Raj and Mala

CONTENTS

Chapter 3   A Computational Approach to the Fusion of Stereopsis and Kineopsis

Amar Mitiche

Chapter 4   The Empirical Study of Structure from Motion

Myron L. Braunstein

Contents

CONTRIBUTORS

Bennett I. Bertenthal *(Chapter 9)*
Department of Psychology, University of Virginia, Gilmer Hall,
Charlottesville, Virginia 22903-2477

Steven D. Blostein *(Chapter 10)*
Coordinated Science Laboratory, College of Engineering, University
of Illinois at Urbana-Champaign,
Urbana, Illinois 61801-3082

Myron L. Braunstein *(Chapter 4)*
Cognitive Sciences Department, School of Social Science, University
of California, Irvine,
Irvine, California 92717

W. Enkelmann *(Chapter 6)*
Fraunhofer-Institut fur Informations- und Datenverarbeitung (IITB),
Sebastian-Kneipp-Str. 12-14, 7500 Karlsruhe 1,
Federal Republic of Germany

David J. Fleet *(Chapter 11)*
Department of Computer Science, University of Toronto,
Toronto, Ontario, Canada M5S 1A4

Thomas S. Huang *(Chapter 10)*
Coordinated Science Laboratory, College of Engineering, University
of Illinois at Urbana-Champaign,
Urbana, Illinois 61801-3082

Michael Jenkin *(Chapter 8)*
Dept. of Computer Science, Ross Building , York University,
4700 Keele St. W.,
Downsview, Ontario, Canada

Allan D. Jepson *(Chapter 11)*
Department of Computer Science, University of Toronto,
Toronto, Ontario, Canada M5S 1A4
Also with The Canadian Institute for Advanced Research.

Joseph K. Kearney   *(Chapter 1)*
Department of Computer Science, University of Iowa,
Iowa City, Iowa 52242

Paul A. Kolers   *(Chapter 8)*
Department of Computer Science, University of Toronto,
Toronto, Ontario, Canada M5S 1A4

R. Kories   *(Chapter 6)*
Fraunhofer-Institut für Informations- und Datenverarbeitung (IITB),
Sebastian-Kneipp-Str. 12-14, 7500 Karlsruhe 1,
Federal Republic of Germany

Amar Mitiche   *(Chapter 3)*
INRS-Telecommunications, 3, place du Commerce, Ile-des-Soeurs,
Verdun, P. Q., Canada H3E 1H6

H.-H. Nagel   *(Chapter 6)*
Fraunhofer-Institut für Informations- und Datenverarbeitung (IITB),
Sebastian-Kneipp-Str. 12-14, 7500 Karlsruhe 1,
Federal Republic of Germany

Keith Price   *(Chapter 5)*
Intelligent Systems Group, University of Southern California,
Los Angeles, California 90089-0273

Dennis R. Proffitt   *(Chapter 9)*
Department of Psychology, University of Virginia, Gilmer Hall,
Charlottesville, Virginia 22903-2477

Brian G. Schunck   *(Chapter 2)*
Computer Vision Research Laboratory, Department of Electrical
Engineering and Computer Science, The University of Michigan,
Ann Arbor, Michigan 48109-2122

Hormoz Shariat   *(Chapter 5)*
Intelligent Systems Group, University of Southern California,
Los Angeles, California 90089-0273

William B. Thompson   *(Chapter 1)*
Computer Science Department, University of Minnesota,
Minneapolis, Minnesota 55455

John K. Tsotsos   *(Chapter 11)*
    Department of Computer Science, University of Toronto,
    Toronto, Ontario, Canada M5S 1A4
    Also with The Canadian Institute for Advanced Research.

Shimon Ullman   *(Chapter 7)*
    Department of Psychology and Artificial Intelligence Laboratory,
    Massachusetts Institute of Technology,
    545 Technology Square, Cambridge, Massachusetts 02139

G. Zimmermann   *(Chapter 6)*
    Fraunhofer-Institut fur Informations- und Datenverarbeitung (IITB),
    Sebastian-Kneipp-Str. 12-14, 7500 Karlsruhe 1,
    Federal Republic of Germany

PREFACE

The physical processes which initiate and maintain motion have been a major concern of serious investigation throughout the evolution of scientific thought. As early as the fifth century B.C. questions regarding motion were presented as touchstones for the most fundamental concepts about existence. Such wide ranging philosophical issues are beyond the scope of this book, however, consider the *paradox* of the flying arrow attributed to Zeno of Elea: An arrow is shot from point A to point B requiring a sequence of time instants to traverse the distance. Now, for any time instant, T_i, of the sequence the arrow is at a position, P_i, and at T_{i+1} the arrow is at P_{i+1}, with $P_i \neq P_{i+1}$. Clearly, each T_i must be a singular time unit at which the arrow is at rest at P_i because if the arrow were moving during T_i there would be a further sequence, T_{i_k}, of time instants required for the arrow to traverse the smaller distance. Now, regardless of the level to which this recursive argument is applied, one is left with the flight of the arrow comprising a sequence of positions at which the arrow is at rest.

The original intent of presenting this paradox has been interpreted to be as an argument against the possibility of individuated objects moving in space. One of the strengths of the argument is that the more one attempts to break the situation down for careful analysis the more one is lead to the given conclusion. Bertrand Russell states the solution as follows: " $\cdots$ but in fact there is no *next* position and no *next* moment, and when this is imaginatively realized, the difficulty is seen to disappear."[1]

Instead of the processes of physical motion, the contributions in this book address the robot and human processes involved in the *perception* of motion from visual information. However, one can trace the implications

---

[1] Russell, B., (1970) 'The problem of infinity considered historically,' in **Zeno's Paradoxes**, W.C. Salmon (ed.), Bobbs-Merrill Co., pp. 51.

of those concerns in many of the issues presented here. For instance, for a robot sensor system in the context of flying arrow stimulus, the sequence of singular time instants is explicit in the form of the visual input -- a sequence of images. If the system expects to detect individual objects, tokens of object parts or just features of interest in each image of the sequence, then the system is faced with the *correspondence* problem.

The majority of the techniques, referred to as 'feature based methods', presume that the correspondence problem is solved. For examples, see the methods presented in Chapters 3, 5, 6, 7, and 10, and the discussions in Chapters 2, 4, 8, and 9. The problem is that for a feature detected in two or more images of the sequence, each of the instances must be recognized as being indicative of the same three-dimensional feature in the sensed environment.

For human perception it is often observed that the person must maintain the perceptual identity of the feature throughout the stimulus interval. Though many researchers have proposed mechanisms by which the human perceptual system might establish and then maintain such perceptual identity, there remains no general agreement. Chapter 8 is fully devoted to this issue and provides evidence that the computational methods which have been found effective for robot systems are not consistent with human behavior. Alternative computational approaches are then suggested.

Chapter 9 investigates further these issues in the context of point-light displays. Such stimuli have been used historically in attempts to isolate the motion interpretation processes. The isolation is presumed to derive from each feature, i.e., point-light, having as its only attribute spatial location and the variation of that location over a time interval. Even with such reduced stimuli, the experiments described in Chapter 9 indicate that complex attributes, e.g., relative/common motion aggregation and perceiver familiarity, play an important role in human perception.

It is this interaction between psychologists interested in human motion perception and researchers primarily interested in computational processes for motion understanding that forms the major strength of this book. The interaction also can be seen clearly in the fact that Chapters 1, 2, 3 and 6 present methods for optical flow. The interest in optical flow can be traced to the theories of the psychologist, J.J. Gibson. In part, the theories claimed that motion information is a primary cue in perception possibly providing the basis for perceptual identity, i.e., correspondence, through time.

The introduction of the image flow equation, see Chapter 2, provided a computational method to derive the image flow as a first approximation to optical flow. The limitations of image flow formulation were immediately recognized and additional constraints proposed to supplement the formulation. Chapters 1 and 2 discuss methods to control the additional constraints. Chapter 3 describes how the problem can be formulated to make use of the additional information in a stereoscopic visual sensor.

The motion information present in optical flow has been suggested as a major cue for establishing and maintaining the perceptual identity, however, severe problems still exist with the optical flow formulation. As is once again made clear in Chapters 1 and 2, the vector fields derived from image flow are not necessarily the optical flow vector fields. Often it is difficult, and in many cases impossible, to derive the correct optical flow from the image flow information. In particular, additional constraints are required when the composition of the three-dimensional environment creates occlusion[2] in the projected images.

The problems created by occlusion in the projection process are difficult to escape in any 'real' environment, particularly over extended time intervals. Even for instantaneous or short-time-interval sampling the image flow calculations may require some indication of the location of occlusion boundaries, i.e., 'jump boundaries' in depth. Unfortunately this presumes segmentation information for the image flow calculation, while

---

[2] A. Mitiche, among others, has pointed out that 'occultation' is a better suited term.

the image flow is to be used to establish and maintain perceptual identity. So a form of the very information one intends to derive is presumed in the calculation.

The feature based methods discussed earlier are also extremely sensitive to occlusion problems. The most obvious problem created by occlusion for feature based methods is that of features or tokens that occur only in disjoint subsequences of the overall image sequence. Maintaining a perceptual identity for features while they are occluded and establishing a perceptual identity for features that are not occluded only briefly is extremely difficult without strong assumptions about the types of features possible and their possible motions. The effects of occlusion on human perception are discussed in detail in Chapter 4 and again in Chapter 9.

In concluding this preface let us relate another *paradox* attributed to Zeno of Elea: A race is to be held between Achilles, the fleetest of the Greeks, and a tortoise. To be fair the tortoise is given a head start, so that at time, T_0, Achilles is at position, P_0, and the tortoise is at P_1. At some time, T_1, Achilles will arrive at P_1, but during that same interval the tortoise will have moved to position, P_2, with $P_2 \neq P_1$. Now, there will also be a T_2 associated with Achilles arriving at P_2 and the tortoise arriving at P_3, then a T_3, and so on. This construction specifies an infinitely long sequence of times, each with Achilles at the previous position of the tortoise, proving that Achilles, regardless of his speed, will never catch the tortoise.

The apparent intent of this *paradox* is to argue against the possibility of physical motion, however, it also questions the divisibility of time into singular instants. For human perception of motion the issue of the divisibility of time is evidenced in the contrast developed between continuous stimuli (often called *real* motion) and discrete stimuli (often called *apparent* motion). The implication of this contrast in human perception is to question the role of *correspondence* in computational processes.

The spatio-temporal filters presented in Chapter 11 attempt to avoid the correspondence problem by 'spatializing time', that is, by treating time as just a third dimension added to the two image dimensions. As presently configured, however, the filters have difficulties with multiple moving objects. Those difficulties are analogous to the occlusion problems for image flow in that segmentation information is required for effective application of the filters.

The apparent circularity in information dependence that occurs in several different approaches to the perception of motion, for both human and robot systems, indicates that there must be at least two distinct phases in motion understanding. The first is the establishment of perceptual identities in the visual information. The second is the maintenance of perceptual identity over extended time intervals, including occlusion subintervals. Both phases are intriguing because they require the integration of 'high' level constraints with 'low' level operations.

It is also clear that any feature based method will require a solution to the fundamental problem of correspondence, particularly when there is a possibility of occlusion. The optical flow approaches attempt to avoid the correspondence problem, but are now faced with the problems of determining effective constraints for the image flow equation and controlling the application of those constraints. In addition, the latter approaches must be formulated to provide mathematically stable computational solutions.

We hope that the contributions in this book will provide a basis for future work in human perception and robot sensing, and a basis for cross-fertilization between these two areas.

W.N. Martin
J.K. Aggarwal

ACKNOWLEDGEMENTS

The editors would like to thank all of the contributors for their continued efforts over the rather extended preparation of this book. For the text formatting we are greatly in debt to Harsha Pelimuhandiram for his unflagging effort at the keyboard. Our efforts were also supported in part by the National Science Foundation through grant DCR–8517583.

MOTION UNDERSTANDING
Robot and Human Vision

CHAPTER 1

Bounding Constraint Propagation
for Optical Flow Estimation

Joseph K. Kearney
William B. Thompson

1.1 Introduction

The velocity field that represents the motion of object points across an image is called the optical flow field. Optical flow results from relative motion between a camera and objects in the scene. One class of techniques for the estimation of optical flow utilizes a relationship between the motion of surfaces and the derivatives of image brightness (Limb and Murphy, 1975; Cafforio and Rocca, 1976; Fennema and Thompson, 1979; Netravali and Robbins, 1979; Schalkoff, 1979; Lucas and Kanade, 1981; Schunck and Horn, 1981; Thompson and Barnard, 1981; and Schalkoff and McVey, 1982). The major difficulty with gradient-based methods is their sensitivity to conditions commonly encountered in real imagery. Highly textured regions, motion boundaries, and depth discontinuities can all be troublesome for gradient-based methods. Fortunately, the areas characterized by these difficult conditions are usually small and localized.

These conditions are especially problematic for methods that operate globally (Horn and Schunck, 1981; Schunck and Horn, 1981). The global method uses numerical relaxation to find the smoothest velocity field consistent with the data. The solution simultaneously minimizes local variation in optical flow and deviation from the gradient constraint. However, abrupt changes in optical flow are common in real images. Optical flow may vary sharply across occlusion edges and at the boundaries of moving objects. The inappropriate enforcement of the smoothness constraint across discontinuities in optical flow can lead to large estimation errors. Even though the error prone regions are sparsely distributed, the global method can propagate inappropriate constraints over large regions causing widespread estimation errors.

Measurement errors in the brightness gradients will also cause errors in optical flow estimates. Frequently, the temporal brightness gradient will be grossly misestimated. Global enforcement of the smoothness constraint can propagate these errors over large areas leading to poor optical flow estimates throughout the image.

In this paper we examine how estimates of the accuracy of optical flow estimates can be used to prevent error propagation. A method is introduced whereby the influence of a point on its neighbor is proportional to the judged correctness of the information to be shared. In this way the mutual constraint of neighbors is controlled and the propagation of errors limited.

1.2 The Gradient Constraint Equation

The gradient constraint equation relates velocity on the image (u,v) and the image brightness function $I(x,y,t)$. The common assumption of gradient-based techniques is that the observed brightness (intensity on the image plane) of any object point is constant over time. Consequently, any change in intensity at a point on the image must be due to motion. Relative motion between an object and a camera will cause the position of a point on the image located at (x,y) at time t to change position on the image over a time interval δt. By the constant brightness assumption, the intensity of the object point will be the same in images sampled at times t and $t+\delta t$. The constant brightness assumption can be formally stated as

$$I(x,y,t) = I(x+\delta x, y+\delta y, t+\delta t). \tag{1}$$

Expanding the image brightness function in a Taylor's series around the point (x,y,t) we obtain

$$I(x+\delta x, y+\delta y, t+\delta t) = I(x,y,t) + \frac{\partial I}{\partial x}\,\delta x + \frac{\partial I}{\partial y}\,\delta y + \frac{\partial I}{\partial t}\,\delta t + \{h.o.t.\}. \tag{2}$$

A series of simple operations leads to the gradient constraint equation:

$$0 = I_x u + I_y v + I_t \tag{3}$$

where

$$I_x = \frac{\partial I}{\partial x}, \quad I_y = \frac{\partial I}{\partial y}, \quad I_t = \frac{\partial I}{\partial t}.$$

A detailed derivation is given in (Horn and Schunck, 1981).

1.3 Gradient-Based Algorithms

The gradient constraint equation does not by itself provide a means for calculating optical flow. The equation only constrains the values of u and v to lie on a line when plotted in flow coordinates.

The gradient constraint is usually coupled with an assumption that nearby points move in a like manner to arrive at algorithms which solve for optical flow. Groups of constraint equations are used to collectively constrain optical flow at a pixel. Horn and Schunck developed a method which globally minimizes an error function based upon the gradient constraint and the local variation of optical flow (Horn and Schunck, 1981). Another approach that has been widely investigated operates locally by solving a set of constraint lines from a small neighborhood as a system of linear equations (Netravali and Robbins, 1979; Schalkoff, 1979; Lucas and Kanade, 1981; Thompson and Barnard, 1981; Kearney, Thompson, and Boley, 1982; and Schalkoff and McVey, 1982; Kearney, 1983).

The local and global methods rely on a similar assumption of smoothness in the optical flow field. Both methods require that flow vary slowly across the image. The locally constructed system of constraint equations is solved as if optical flow is constant over the neighborhood from which the constraint lines are collected. When optical flow is not constant, the local method can provide a good approximation where flow varies slowly over small neighborhoods. The global method seeks a solution which minimizes local variation in flow. The important difference between methods of local and global optimization is not the constraint that they place on the scene but the computational method that they use to apply the constraint. There are contrasting aspects in the performance of the two approaches that are directly related to the difference in the scope of interactions across the image.

The narrow focus of local methods leads to a major problem over portions of the image. The local optimization scheme has difficulty in regions where the spatial gradients change slowly. In these regions the gradient constraint lines will be nearly parallel; consequently, the linear system will tend to be ill-conditioned. Unavoidable measurement errors in the gradient estimates will be magnified in the estimated value of flow. The problem arises because local information is insufficient for estimating flow in regions where the orientation of the spatial gradient is nearly constant. The group of constraint lines provide essentially the same constraint on optical flow. These problems are reduced in the global optimization method. Information is propagated over the image, so regions which have insufficient local constraints will benefit from the estimates at surrounding regions.

The local and global methods share a common weakness. Where flow changes sharply estimates will be very inaccurate. The affect of these errors is limited by the neighborhood size in the local method. In contrast, global methods may propagate these errors throughout the image. While the global sharing of information is beneficial for constraint sharing, it is detrimental with respect to error propagation. Without some capability to contain interactions to separate regions that satisfy the smoothness assumption, global optimization methods are practically useless for most real imagery.

1.4 Coping with Smoothness Violations

The difficulty encountered with global methods is that if information is inappropriately combined across discontinuities in the flow field large estimation errors can result. These errors tend to propagate throughout the flow field, even though the problematic regions may represent only a small portion of the image.

1.4.1 Thresholding for Smoothness

One way to approach the problem is to attempt to explicitly identify the boundaries of regions which internally satisfy the smoothness constraint. Schunck and Horn suggest two heuristics for identifying neighboring constraint equations which differ substantially in their flow value (Schunck and Horn, 1981). Once these points are identified they can be removed from the network. If all of these points are located, the image should naturally be segmented into regions bounded by flow discontinuities. Within these regions the smoothness assumption is expected to hold. The difficulty with this approach is the selection of a threshold for identifying unsmoothness.

Smoothness can not be measured directly so we must measure conditions that are usually associated with large changes in flow. For example, Schunck and Horn argue that pairs of constraint equations that lie across a discontinuity in flow are likely to intersect at large values of flow. They also suggest that the orientation of the constraint lines obtained from points which lie on different surfaces, with different reflectance functions, are likely to differ significantly. They propose a heuristic for identifying points which are likely to lie along a discontinuity in optical flow. If a pair of neighboring constraint equations *either* intersect at a value greater than a threshold or differ in orientation by more than a second threshold, then these points are judged to lie across a discontinuity in flow. We must be a little cautious in applying rules such as this, understanding that they are probabilistic. If the thresholds for the detection of unsmoothness are set too low, we will likely miss parts of the boundary. A high threshold, however, will result in many points being incorrectly identified as lying along flow discontinuities. Removal of many points which do not violate the smoothness constraint can cause a deterioration of the overall system.

The problem is compounded because these conditions for finding flow edges are also likely to occur within regions over which flow varies smoothly. Wherever the orientation of the gradients changes slowly, small amounts of noise can cause pairs of constraint lines to intersect at large, incorrect flow values. This is due to the conditioning problem which plagues local methods. These points should not cause problems for the global approach and should not be removed from the field. Furthermore, it is exactly the variation in the orientation of constraint equations that provides the second constraint necessary to estimate flow. Casual removal of points where the orientation varies sharply can reduce the contribution of the second constraint. The difficulty is that information bearing points are removed and regions which should share constraints may be isolated from each other.

Other methods to segment the image into regions over which flow varies smoothly have been proposed. One approach uses locally derived estimates of optical flow to find motion boundaries. The global method is then used to refine flow estimates within closed regions (Schunck, 1985). Another approach uses brightness edges to identify potential motion boundaries. Constraints are selectively propagated across brightness edges by comparing locally derived estimates obtained with and without information exchange across the boundary (Cornelius and Kanade, 1983).

The difficult problem that must be addressed by all of the segmentation schemes is the determination of a threshold. The threshold must be set relatively high to prevent inappropriate exchange of information across boundaries. This may lead to unnecessary barriers to constraint propagation thereby reducing the advantage of global sharing of information.

1.4.2 Continuous Adaptation to Errors

Smoothness violations need not be treated categorically. The variations of optical flow will occur in continuous gradations as will the resulting estimation errors. We present an approach that accepts the inevitability of errors, some possibly large, and attempts to use knowledge about the reliability of estimates to preserve good information and attenuate the propagation of poor estimates. Instead of attempting to eliminate these errors, we will try to judge the accuracy of flow estimates and propagate estimates in proportion to their accuracy. An advantage of this approach is that we can attempt to minimize the propagation of other types of error as well.

A measure of confidence must be associated with each optical flow estimate. The confidence measure should represent our belief in the accuracy of an estimate. We can incorporate information about conditions which are likely to lead to errors -- indications of unsmoothness, for example -- into the confidence measure. When an estimate has been calculated, we can sometimes identify poor estimates by examining how well constraints have been satisfied. These *a posteriori* judgements of accuracy can also contribute to the confidence measure.

We need to develop a method to compute optical flow which propagates estimates in proportion to their confidence value. The Horn and Schunck technique can be viewed as a process of constrained smoothing. Let $(\bar{u},\bar{v})$ be the average of neighboring estimates of optical flow. The point (u_p,v_p) is defined as a point on the gradient constraint equation which also lies on the perpendicular to the gradient constraint line that passes through $(\bar{u},\bar{v})$ (see Figure 1). The computational method used by Horn and

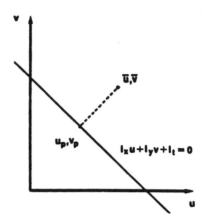

Figure 1. Combining the gradient constraint equation
and the local average of optical flow.

Schunck is equivalent to taking a weighted combination of $(\overline{u},\overline{v})$ and
(u_p,v_p). The weighting is determined by the magnitude of the spatial gradients. The process is performed iteratively over the entire image. It can
be shown that this method will converge to a solution that globally minimizes violations of the gradient and smoothness constraints (Horn and
Schunck, 1981).

We can incorporate accuracy judgements into the global method by
computing the local average motion as the average of neighboring optical
flow estimates weighted by confidence. Since estimates contribute only in
proportion to their confidence, good estimates will tend to propagate more
effectively. A new estimate of optical flow is obtained by combining the
weighted average of local flow estimates and the gradient constraint equation.

Estimating Confidence

The success of this method depends on our ability to judge the accuracy of optical flow estimates. Our implementation assigns a confidence to
a new estimate on the basis of

(a)   the confidence placed in the local average,

(b)   an estimate of smoothness,

(c)   the estimated error in the gradient constraint equation, and

(d)   an *a posteriori* bound on the error in the new flow estimate.

The confidence in the *local average* can be judged by the mean confidence of the estimates which contribute to the average flow estimate. Estimates should contribute to the mean confidence statistic in the same proportion as they contribute to the average optical flow estimate.

Violations in the *smoothness assumption* are detected in two ways. The agreement between the local average flow and the gradient constraint line provides a simple index of errors that can result from a lack of smoothness. The gradient constraint line should pass through the true value of optical flow; and where optical flow varies smoothly, $(\bar{u},\bar{v})$ should lie near the true value of optical flow. If $(\bar{u},\bar{v})$ is well separated from the gradient constraint line, then it can be inferred that one or the other is likely to be in error. Another more direct way to identify unsmoothness is to compute the variance among neighboring estimates of optical flow. Large local variations in optical flow may be due to a lack of smoothness or estimations errors. In either case the new estimate based upon these values is likely to be incorrect. Note that the confidence in the neighboring estimates must be considered in the calculation of the variance. Flow estimates should contribute to the variance calculation in proportion to their contribution to the newly estimated flow value.

The *gradient constraint* equation will be in error due to inexactness in the measurements of the gradients. Each gradient estimate will contain a component of random error due to noise in the imaging process. Gradients will also contain a component of error that is systematically related to the linearity of the brightness function. The gradients are usually approximated by the change in brightness over some small interval of the sampled image function. The error in the gradient approximation is proportional to the nonlinearity of the brightness function over the sampling interval.

Large errors in the estimate of the temporal gradient can arise in regions where the spatial brightness function contains nonlinearities and optical flow is significant (Lucas and Kanade, 1981; Kearney, Thompson, and Boley, 1982; Kearney, 1983). The change in the spatial gradient at a point over time provides a rough index of this condition. The magnitude of the change in the spatial gradient can be used to detect locations where the gradient constraint equation is likely to be grossly incorrect.

The accuracy of flow estimates can be gauged *a posteriori* by examining the degree to which predicted motions fit the actual image sequence. The optical flow field predicts the motion of surface points from frame to frame of an image sequence. We can test the accuracy of flow estimates by comparing predicted appearance of surface regions with that actually observed. The correlation of the brightness function over small areas around the expected locations of a point in successive temporal samples might be used as a measure of success.

A simple *a posteriori* error bound can be obtained by referring to a more general form of the gradient constraint equation (Kearney, 1983; Paquin and Dubois, 1983). Consider the image sequence that samples the three-dimensional image function. We actually estimate the displacement of a point between successive samples of the image sequence. If velocity is constant, then the displacement observed on the image, (d_x, d_y), over the time interval Δt is $(u\Delta t, v\Delta t)$. Let $\mathbf{d}$ be a displacement vector in 3-dimensional x,y,t-space. Let $\hat{\mathbf{d}}$ be an estimate of $\mathbf{d}$. Given a displacement estimate

$$\hat{\mathbf{d}} \equiv \begin{bmatrix} \hat{d}_x \\ \hat{d}_y \\ \Delta t \end{bmatrix} = \begin{bmatrix} \hat{u}\Delta t \\ \hat{v}\Delta t \\ \Delta t \end{bmatrix} = \begin{bmatrix} x\text{-component of displacement} \\ y\text{-component of displacement} \\ t\text{-component of displacement} \end{bmatrix} \qquad (4)$$

we can estimate optical flow by $(\hat{u}, \hat{v})$.

The vector $\dfrac{\hat{\mathbf{d}}}{\|\hat{\mathbf{d}}\|}$ is a unit vector in the direction of the estimated displacement. The gradient of I in this direction is

$$I_{\hat{\mathbf{d}}} = \frac{\hat{\mathbf{d}}^T}{\|\hat{\mathbf{d}}\|}\begin{bmatrix} I_x \\ I_y \\ I_t \end{bmatrix} = \frac{1}{\|\hat{\mathbf{d}}\|}(I_x\hat{u}+I_y\hat{v}+I_t)\Delta t \qquad (5)$$

$$= \frac{1}{\|\hat{\mathbf{d}}\|}(I_x\hat{u}+I_y\hat{v}-I_xu-I_yv)\Delta t \quad \text{(using (3))} \qquad (6)$$

$$= \frac{I_x(\delta d_x)+I_y(\delta d_y)}{\|\hat{\mathbf{d}}\|} \qquad (7)$$

where $\delta d_x = \hat{d}_x - d_x$ and $\delta d_y = \hat{d}_y - d_y$ are the errors in the components of the displacement estimate. Finally, by rearranging (7) we arrive at a bound on the error in the displacement estimate:

$$\|(\delta d_x, \delta d_y)\| \geq \frac{\|\hat{\mathbf{d}}\| I_{\hat{\mathbf{d}}}}{\|(I_x, I_y)\|}. \qquad (8)$$

The righthand side of (8) can be estimated from the image sequence using estimates of the spatial gradients and the approximation

$$\|\hat{\mathbf{d}}\| I_{\hat{\mathbf{d}}} \approx I(x+\hat{u}\Delta t, y+\hat{v}\Delta t, t_1+\Delta t) - I(x,y,t_1). \qquad (9)$$

The error bound in (8) can be easily calculated for each displacement estimate.

Combining Partial Estimates of Confidence

In this chapter several ways to evaluate the correctness of optical flow estimates have been introduced. A single value that represents our confidence in an estimate must be derived from many sources of information about the likelihood of error. The confidence value will determine the degree to which the optical flow estimate at a given point influences motion estimation at other locations. It is computationally convenient to let confidence vary between 0 (judged inaccurate) to 1 (judged accurate). This allows the weighted average of flow and the confidence in the average to be calculated simply. The conditions presented to detect errors -- symptoms of unsmoothness, the change of spatial gradient, and the *a posteriori* error bound -- can all be interpreted as measurements of error, each of which ranges from 0 upward. These can be converted into confidence-like statistics by adding one and taking the inverse. A set of confidence measurements which vary from 0 to 1 is obtained. We found that it is best to treat the elements as independent and combine the set of confidence measures multiplicatively. The product rule possesses the characteristics we desire -- it is sensitive to the many conditions which can cause errors in the global system and it is conveniently interpreted. We have empirically examined a number of combination rules and found that the results were not highly sensitive to the particular rule for combining error measurements.

1.5 Results

Quantitative evaluation of techniques that estimate optical flow is difficult. The criteria by which we judge performance should depend upon the requirements of the interpretation processes that will use the flow field. And as yet, the interpretation processes are not well understood. Evaluation of gradient-based methods is further complicated by the number of parameters that can be adjusted and the enhancements that can be added to improve performance. For example, gradient-based methods require some amount of smoothing to remove nonlinearities in the brightness function that cause measurement errors in the gradient estimates. The amount of smoothing and the method by which smoothing is accomplished vary from implementation to implementation. Methods of iterative registration and coarse-to-fine analysis have also been proposed (Lucas and Kanade, 1981). The choice of parameters and enhancements can significantly affect performance. It has been suggested that the selection be adaptive to conditions in the image and the requirements of the interpretation task (Kearney, 1983).

We have implemented several versions of local and global gradient-based methods with enhancements. We present results from a technique that in our judgement performed the best with a variety of scenes. The method is based upon the global gradient method with propagation by confidence as described above. While quantitative results are not given, the data demonstrate the viability of the approach.

Flow estimates were computed by combining the weighted average of flow estimates and the gradient constraint equation. The local average was computed over a 5x5 neighborhood, centered on the point to be estimated. The gradient constraint equation and the local average are combined as in (Horn and Schunck, 1981). Estimates are globally propagated by iterating the local estimation process on a single frame pair. The estimates we present were the result of 16 iterations of the averaging process.

Before the global method can proceed, the optical flow field must be assigned initial values. The technique could begin with a field of zero flow vectors assigned some default level of confidence. If the technique is to be used over a sequence of more than two images, the results of the previous iteration can be used as an initial approximation of the flow field. The method can also be seeded with estimates obtained elsewhere (Yachida, 1983). Our implementation first performed a correlation matching procedure as in (Moravec, 1980). The matching results were assigned a confidence based upon the degree of correlation at the determined match. All other points in the flow field were assigned a zero flow vector with zero confidence.

Our method also incorporates iterative registration as suggested in (Lucas and Kanade, 1981). Previous estimates of optical flow are refined on successive iterations of the global averaging process. In effect, we register small patches of the image and solve for the error in our registration. An improvement results from a likely reduction in the measurement error in the temporal gradient (Kearney, 1983).

The method was tested with the two image pairs presented in Figures 2 and 3.

Figure 2. The flyover image sequence.

The first sequence simulates a view from an aircraft flying over a city. The scene is actually a model of downtown Minneapolis. (This picture originally appeared in (Barnard, 1979).) The second sequence was taken with a fixed camera viewing two toy trains moving towards one another.

Optical flow fields obtained with the global technique are shown in Figures 4 and 5 for the flyover and moving train scenes, respectively. The method associates a confidence value with each flow vector. For each figure, a threshold was set to determine which vectors from the flow field would be displayed. Only 20% of the flow vectors with confidence values that exceeded the threshold are displayed in each figure. The flyover flow field is presented using an intermediate value for the threshold on confidence (Figure 4). The flow field produced with the train sequence is displayed using three different thresholds on confidence in Figures 5.a, 5.b, and 5.c, ordered from lowest to highest threshold.

Figure 3. The moving trains image sequence.

Figure 4. The optical flow field for the flyover sequence.

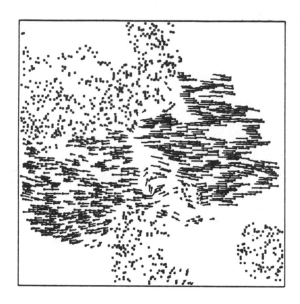

Figure 5a. The optical flow field for the moving trains sequence.
Low confidence threshold.

The confidence estimates provide a reasonable index of the accuracy of optical flow estimates. A sparse sampling of accurate estimates exceeds the high confidence threshold. When the threshold is lowered, more dense fields are obtained with correspondingly more errors evident.

Note that the areas where very few vectors are displayed. Optical flow is poorly estimated in these regions and the resulting values of confidence are low. The problematic regions usually fit one or more of the following characterizations:

(1)   Regions with largely homogeneous intensity values containing little local information for optical flow and not easily assigned a flow value from surrounding regions.

(2)   Highly textured regions which are moving, leading to poor estimates of the temporal gradient.

(3)   Regions which contain large discontinuities in the flow field, thus violating the smoothness assumption.

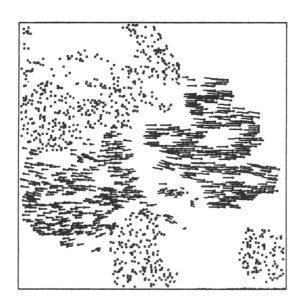

Figure 5b. The optical flow field for the moving trains sequence.
Intermediate confidence threshold.

1.6 Discussion

The results demonstrate the feasibility of measuring the quality of optical flow estimates. Gradient-based methods are susceptible to a variety of problems and tend to produce very poor estimates in troublesome areas of the image. Without accurate estimates of confidence, good flow estimates can not be distinguished from bad and gradient-based techniques are of little use. With accurate confidence estimates, poor optical flow estimates can be filtered from the field and information can be propagated from areas of high information content into areas of low information content.

The problem of managing constraint propagation arises in a number of other contexts. Recently, many vision problems have been formulated as minimization problems which globally enforce a local smoothness constraint. For example, the estimation of shape from shading information can be approached in this way (Ikeuchi and Horn, 1979). However, real

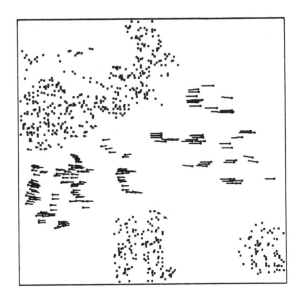

Figure 5c. The optical flow field for the moving trains sequence.
High confidence threshold.

images usually contain abrupt changes in depth, motion, and surface orientation. If the smoothness constraint is propagated across discontinuities of the intrinsic property to be estimated, widespread errors can result.

The approach presented here may provide a means of coping with constraint propagation for problems other than motion estimation. If the correctness of estimates can be judged, then it is reasonable to consider regulating the propagation of constraints by confidence measures.

Acknowledgements

This work was supported by the Air Force Office of Scientific Research contract, F49620-83-0140, and the National Science Foundation Grant, MCS-81-05215.

References

Barnard, S.T., (1979) 'The image correspondence problem,' Computer Science Department, University of Minnesota, PhD. Dissertation.

Cafforio, C., and Rocca, F., (1976) 'Methods for measuring small displacements of television images,' *IEEE Trans. on Information Theory* vol. IT-22, pp. 573-579.

Cornelius, N., and Kanade, T., (1983) 'Adapting optical-flow to measure object motion in reflectance and X-ray image sequences,' *ACM Interdisciplinary Workshop on Motion: Representation and Perception*, Toronto, Canada.

Fennema, C.L., and Thompson, W.B., (1979) 'Velocity determination in scenes containing several moving objects,' *Computer Graphics and Image Processing*, vol. 9, pp. 301-315.

Horn, B.K.P., and Schunck, B., (1981) 'Determining optical flow,' *Artificial Intelligence*, vol. 17, pp. 185-203.

Ikeuchi, K., and Horn, B.K.P., (1979) 'An application of the photometric stereo method,' *Proc. 6th Int. Joint Conf. on Artificial Intelligence*, pp. 413-415.

Kearney, J.K., (1983) 'Gradient-based estimation of optical flow,' University of Minnesota, PhD. Dissertation.

Kearney, J.K., Thompson, W.B., and Boley, D.L., (1982) 'Gradient-based estimation of disparity,' *Proc. IEEE Conf. on Pattern Recognition and Image Processing*.

Limb, J.O., and Murphy, J.A., (1975) 'Estimating the velocity of moving images in television signals,' *Computer Graphics and*

Image Processing, vol. 4, pp. 311-327.

Lucas, B.D., and Kanade, T., (1981) 'An iterative image registration technique with an application to stereo vision,' *Proc. of the 5th Joint Conf. on Artificial Intelligence,* pp. 674-679.

Moravec, H.P., (1980) 'Obstacle avoidance and navigation in the real world by a seeing robot rover,' Stanford University, PhD. Dissertation.

Netravali, A.N., and Robbins, J.D., (1979) 'Motion-compensated television coding: part I,' *The Bell System Tech. J.,* vol. 58, no. 3.

Paquin, R., and Dubois, E., (1983) 'A spatio-temporal gradient method for estimating the displacement field in time-varying imagery,' *Computer Vision, Graphics and Image Processing,* vol. 21, no. 2, pp. 205-221.

Schalkoff, R.J., (1979) 'Algorithms for a real-time automatic video tracking system,' University of Virginia, PhD. Dissertation.

Schalkoff, R.J., and McVey, E.S., (1982) 'A model and tracking algorithm for a class of video targets,' *IEEE Trans. on Pattern Analysis and Machine Intelligence,* vol. PAMI 1-4, no. 1, pp. 2-10.

Schunck, B.G., (1985) 'Image flow: Fundamentals and future research,' *Proc. IEEE Conf. on Pattern Recognition and Image Processing,* pp. 560-571.

Schunck, B.G., and Horn, B.K.P., (1981) 'Constraints on optical flow,' *Proc. IEEE Conf. on Pattern Recognition and Image Processing,* pp. 205-210.

Thompson, W.B., and Barnard, S.T., (1981) 'Low-level estimation and interpretation of visual motion,' *Computer*, vol. 14, no. 8, pp. 20-28.

Yachida, M., (1983) 'Determining velocity maps by spatio-temporal neighborhoods from image sequences,' *Computer Vision, Graphics and Image Processing*, vol. 21, no. 2, pp. 262-279.

CHAPTER 2

Image Flow: Fundamentals and Algorithms

Brian G. Schunck

2.1 Introduction

This chapter describes work toward understanding the fundamentals of image flow and presents algorithms for estimating the image flow field. Image flow is the velocity field in the image plane that arises due to the projection of moving patterns in the scene onto the image plane. The motion of patterns in the image plane may be due to the motion of the observer, the motion of objects in the scene, or both. The motion may also be apparent motion where a change in the image between frames gives the illusion of motion. The image flow field can be used to solve important vision problems provided that it can be accurately and reliably computed. Potential applications are discussed in Section 2.1.2.

This chapter is concerned primarily with the analysis and algorithm development for image flows that may contain discontinuities in either the image irradiance or the image flow velocity field. Image flows with continuous translation of smooth patterns of image irradiance across the image plane are less interesting since they do not exhibit the discontinuities that occur in real images and that allow the image flow velocity field to provide valuable information for motion segmentation.

2.1.1 Background

Before proceeding with image flow, two related areas of motion research must be mentioned: difference picture methods and matching. Difference picture and matching methods lie at opposite ends of a spectrum of visual motion topics from local difference picture methods that only involve single pixels to global feature matching methods that span the entire image. Image flow is a short-range visual process that is not as global as matching methods, but not as local as difference picture methods.

A difference picture is a pixel-wise difference between successive frames in an image sequence (Jain, Militzer, and Nagel, 1977; Jain and Nagel, 1979; Jain, Martin, and Aggarwal, 1979; Yalamanchili, Martin, and Aggarwal, 1982; and Jain, 1983). The differences can be thresholded to produce a binary motion image. Since difference picture methods do not take the direction or speed of motion into account, the methods are not easily applied when the background is moving and cannot differentiate between different objects moving with different velocities. The problem with difference picture methods is that they are too local; they do not combine spatial and temporal changes over small neighborhoods to use the information in the velocity field.

Matching methods lie at the other end of the local to global spectrum. Ullman (1979b) studied algorithms for matching the components of objects between two successive frames of a scene containing moving objects. Matching leads to an optimization problem which can be solved by a simple network (Ullman, 1979c). The drawback to this approach for image flow is that identifiable features must be derived from the image before matching can be tried. The spatial range over which image flow is computed is short in comparison to the size of the region in which identifiable features exist. This chapter is concerned with designing an image flow algorithm that is analogous to the short-range motion process. Prior detection or classification of features in the image sequence is not performed.

Work in psychology indicates that there are two motion estimation processes in the human vision system: one process is a global or long-range process that matches image features between successive frames of an image sequence, the other process is local or short-range and does not seem to work on image features that require extensive preprocessing. The work of Braddick (1974) quantifies the distinction between short- and long-range motion processes. The clarity of the motion percept decreases linearly between 5' and 20' and beyond 20' the central square is not perceived as moving coherently. The range over which the clarity of the motion perception deteriorates corresponds closely with the increasing sizes of the channels in the edge detection theory of Marr and Hildreth (1980). As the displacement of the square increases, the movement is contained within the support of fewer channels and the ability to perceive the motion decreases. Above 20', the motion is beyond the support of even the largest channel and the ability to perceive coherent motion is lost. The range over which short-range motion phenomena occur is limited by the size of the largest channel. It can be inferred by analogy with this result in biological vision that in a machine vision system, the maximum displacement for which an image flow estimation algorithm can be applied is limited by the size of the smoothing filter applied to the image or the size of the aperture function if no smoothing is performed. The aperture function is usually approximately equal to the pixel spacing. If no smoothing filter is applied to the image, then the displacement between frames must be no greater than the pixel spacing.

2.1.2 *Applications for Image Flow*

Several potential applications may exist for image flow. Algorithms have been developed to solve important problems in vision processing such as estimating surface structure or egomotion from the image flow velocity field. To avoid confusion, the term optical flow analysis will be used to denote algorithms that use the image flow velocity field for solving a vision problem such as structure from motion or egomotion, while the terms image flow estimation or velocity field estimation will be used to denote algorithms that compute the velocity field in the image plane without further analysis. Historically, the perceived impediment to applying algorithms for optical flow analysis has been the absence of an algorithm for estimating the image flow velocity field from image input. The analysis algorithms, as formulated so far, assume that the image flow velocity field has already been computed and is readily available as an intrinsic image. As will become clear in later sections, computing the image flow velocity field is very difficult. A trend may develop toward reformulating optical flow analysis algorithms to avoid the need to compute the image flow velocity field as an intermediate step. Nevertheless, the image flow estimation algorithm presented in this chapter in Section 2.5 is useful for scene segmentation, even if it may not produce a velocity field estimate with sufficient accuracy for further analysis. If the image flow velocity field produced by the algorithm can be used as input to algorithms for optical flow analysis, then this will be an added benefit. The principle accomplishments of this work are (1) the development of an equation for image flow that is appropriate for use with image irradiance or velocity field discontinuities, and (2) the development of an algorithm for image flow estimation that works when there are image irradiance or velocity field discontinuities and preserves the sharp motion boundaries. Perhaps this research will inspire approaches to developing algorithms for motion analysis, such as structure from motion and egomotion, when there are discontinuities in the velocity field or image irradiance.

Structure from Motion

It has been shown that the motion parallax field can be decomposed into the components of curl, divergence, and deformation (Koenderink and van Doorn, 1975 and 1976). The result provides no clue as to how the image flow field may be computed, although the decomposition of the velocity field into these components may be useful in algorithms for estimating surface structure from image flow.

Clocksin (1978) investigated the possibility of computing the surface orientation of objects in a scene from the image flow caused by observer motion. Spherical coordinates were used; the axis of the sphere coincided with the direction of motion of the observer. The direction of local image flow is always oriented along the meridian lines of the sphere and the speed of flow is given by

$$\frac{d\beta}{dt} = \frac{S \sin \beta}{r}, \tag{1}$$

where S is the speed of motion, β is the angle of latitude on the sphere, and r is the distance from the observer, i.e., the center of the sphere, to the object. Surface orientation in coordinates aligned with the lines of latitude and longitude is given by

$$\tan \sigma = \frac{1}{r}\frac{\partial r}{\partial \beta} \quad \text{and} \quad \tan \tau = \frac{1}{r}\frac{\partial r}{\partial \alpha}, \tag{2}$$

where α and β are the angles of the longitude (meridian) and latitude, respectively. Later work showed that the distance r to the object and the speed of motion S of the observer are unnecessary for computing surface orientation, and analyzed the properties of the discontinuities in the image flow caused by occluding and non-occluding edges (Clocksin, 1980).

Ullman (1979a and 1979b) studied structure from motion determined by token correspondence. The estimation of motion parameters has been well studied by Tsai and Huang (1982) and Tsai, Huang, and Zhu (1982).

A novel approach to motion analysis using pairs of image flow fields was proposed by Waxman and Sinha (1984). Waxman and Ullman (1983), Waxman (1984), and Waxman and Wohn (1984) have advocated the estimation of motion and surface structure through the analysis of contour deformation due to motion.

Hoffman (1980) described an algorithm for computing the surface orientation and angular velocity of a local patch of surface on an object from the spatial derivatives of the velocity and acceleration field of the surface patch projected onto the image plane by orthographic projection. It may not be possible to compute both the velocity and acceleration fields with sufficient accuracy for computing the spatial derivatives of these fields.

Longuet-Higgins and Prazdny (1980) presented equations for determining scene structure and object motion from the motions of features in the image plane. Further work on determining scene structure from the motion of edge features has been presented by Buxton, Buxton, Murray, and Williams (1983). This work is not the same as estimating scene structure from the image flow velocity field: only the normal component of the edge velocity was used. The development avoids requiring the computation of the velocity field as an intermediate step towards scene analysis.

Egomotion and Passive Navigation

Given the image flow field it is possible, in principle, to determine the motion of the observer relative to the observed scene. This application, called egomotion in the perception literature, was studied by Prazdny (1980). In aircraft navigation, this visual process is called passive navigation since it determines the motion of an aircraft without active probing by radar. Passive navigation is a simple version of the general egomotion problem: the terrain is stationary relative to the absolute frame of reference. Horn and Bruss (1983) used the least squares method to obtain the motion of an observer with respect to a fixed scene from the image flow field.

Object Tracking

In object tracking, the goal is to extract from the image sequence an estimate of the motion of objects in the scene (Roach and Aggarwal, 1979; Hirzinger, Landzettel, and Snyder, 1980; Bers, Bohner, and Gerlach, 1980; Snyder, Rajala, and Hirzinger, 1980; Price, Snyder, and Rajala, 1981; and Schalkoff and McVey, 1982). It is frequently assumed that only one object in the scene is moving. The motion estimate can be used to control the camera orientation so that the object is tracked by camera motion or can be used to provide input to another subsystem that tracks the object.

Flinchbaugh and Chandrasekaran (1981) worked on the problem of aggregating the motion of features into coherent motions that could correspond to objects. This work is not usually associated with object tracking, but should be mentioned in this context since the two-step process of estimating feature motion followed by aggregation of feature motions into object motions should lead to better object tracking algorithms that can handle multiple moving objects undergoing complex motions.

Image Compression

The goal of image compression is to reduce the bandwidth required for transmitting a television signal by recoding the image sequence to eliminate the redundancy between frames. The encoding can be more effective if the motion of regions in the scene can be determined so that the image points between which the modulation differences are derived can be chosen to correspond to displaced versions of the same patch of image intensity.

2.1.3 Summary

Section 2.2 will cover simple image flows, which are image flows that do not contain discontinuities. Section 2.3 will extend the derivation of the equation for image flow to cases where there are discontinuities in the image irradiance or the velocity field. Section 2.4 will elaborate on the development of the image flow equation by reformulating the equation into a polar form more appropriate for image flows with discontinuities. Section 2.5 will present an algorithm for estimating the velocity field in the presence of discontinuities and Section 2.6 presents an algorithm for smoothing the velocity field estimate between motion boundaries without blurring the boundaries.

2.2 Simple Image Flows

In this section, the image flow equation for the case of simple flow of image irradiance across the image plane will be examined. Simple image flows are velocity fields in the image plane that do not contain velocity field discontinuities and are derived from images that do not contain image irradiance discontinuities. Early image flow research formulated a simple image flow equation for smooth patterns of image irradiance and smooth velocity fields and developed algorithms for solving simple image flows. Work described in Sections 2.3 and 2.4 extends the equation to image irradiances and velocity fields with discontinuities.

2.2.1 Image Flow Equation for Simple Flows

The image flow constraint equation

$$E_x u + E_y v + E_t = 0, \qquad (3)$$

relates the temporal and spatial changes in the image irradiance, $E(x,y,t)$, at a point, (x, y), in the image plane to the instantaneous velocity, (u, v), at that point in the image. This equation assumes that the pattern of image irradiance translates across the image plane without distortion in the pattern.

To derive the image flow constraint equation, imagine a patch of image irradiance that is displaced by δx in the x-direction and by δy in the y-direction in time δt. The patch of image irradiance is assumed to remain unchanged between displacements so

$$E(x,y,t) = E(x+\delta x, y+\delta y, t+\delta t). \qquad (4)$$

The right hand side can be expanded about the point (x,y,t) using Taylor series

$$E(x+\delta x,y+\delta y,t+\delta t) = E(x,y,t) + \frac{\partial E}{\partial x}\delta x + \frac{\partial E}{\partial y}\delta y + \frac{\partial E}{\partial t}\delta t + \varepsilon, \qquad (5)$$

where ε contains second and higher order terms in δx, δy, and δt. After subtracting $E(x,y,t)$ from both sides and dividing through by δt, the result is

$$\frac{\partial E}{\partial x}\frac{\delta x}{\delta t} + \frac{\partial E}{\partial y}\frac{\delta y}{\delta t} + \frac{\partial E}{\partial t} = O(\delta t), \qquad (6)$$

where $O(\delta t)$ is a term of order δt that includes the first and higher variations of x and y since it is assumed that δx and δy will depend on δt. In the limit as $\delta t \rightarrow 0$, the equation above becomes

$$E_x u + E_y v + E_t = 0. \tag{7}$$

This derivation assumes that the patterns of image irradiance are translating across the image plane without discontinuities in the velocity field. More formally, the derivation assumes that the image irradiance and velocity fields are analytic in some neighborhood of the displacement. The derivation also assumes that there are no discontinuities in the pattern of image irradiance. Discontinuities in $E(x,y,t)$ would lead to δ-functions in the partial derivatives. This derivation is not valid when the image irradiance contains discontinuities or when there are image flow discontinuities due to occlusion.

2.2.2 Algorithms for Simple Image Flows

Algorithms for estimating simple image flows without discontinuities in the image irradiance or velocity field will be summarized in this section. Algorithms developed by optimization over the image or along edges will be presented, and correlation as a motion estimation algorithm will be mentioned. Algorithms developed in the context of image compression will be mentioned. An overview of work in image flow and motion analysis is presented by Nagel (1980).

Image Flow Estimation by Optimization

Horn and Schunck (1981) formulated an optimization problem for estimating image flow. The optimization measure consisted of two terms: a penalty on the deviation of the estimated velocity field from the image flow constraint equation and a penalty on the deviation of the velocity field components from smooth surfaces. The smoothness penalty was the sum of the squares of the magnitude of the gradient of image flow velocity. The optimization criterion was

$$\int \int \left[E_x u + E_y v + E_t \right]^2 + \kappa^2 \left[\left[\frac{\partial u}{\partial x} \right]^2 + \left[\frac{\partial u}{\partial y} \right]^2 + \left[\frac{\partial v}{\partial x} \right]^2 + \left[\frac{\partial v}{\partial y} \right]^2 \right] dxdy, \quad (8)$$

where the parameter κ^2 controls the relative cost of deviations from smoothness and deviations from the motion constraint. By applying the calculus of variations (Courant and Hilbert, 1937, pp. 191-192), a pair of coupled partial differential equations are obtained,

$$\Delta u = \frac{E_x}{\kappa^2} (E_x u + E_y v + E_t)$$

$$(9)$$

$$\Delta v = \frac{E_y}{\kappa^2} (E_x u + E_y v + E_t).$$

An estimate of the image flow velocity field can be obtained by solving this pair of partial differential equations for u and v given the partial derivatives with respect to x, y, and t of the image irradiance, $E(x,y,t)$. Iterative equations have been implemented (Horn and Schunck, 1981) for solving these equations and examples presented of the application of the iterative equations to several synthetic images.

Unfortunately, the algorithm cannot work in cases where there are discontinuities in the velocity field since the smoothness constraint leads to iterative equations that blur the abrupt changes in the velocity field. Since the derivation of the image flow constraint equation is unsound for cases where there are image irradiance or velocity field discontinuities, it is unclear whether the image flow equation itself is valid for situations with discontinuities. This question will be resolved in Section 2.3 and an image flow estimation algorithm that preserves discontinuities in the velocity field will be presented in Section 2.5.

Motion Estimation by Correlation

The visual system of the fly, which is noted for its ability to perceive motion, has been studied (Poggio and Reichardt, 1976; Reichardt and Poggio, 1979; and Buckner, 1976) under the assumption that the motion detection capability could be modelled by a multi-input system with a Volterra series representation (Schetzen, 1980), where the inputs are the image irradiances detected by the eye lens facets. Volterra series can be classified and systems with Volterra series expansions in different classes respond differently to the same stimulus. Experiments to determine the character of the Volterra series for the motion detection system of the fly were formulated and it was concluded that the behavior of the motion detection system can be described as a correlation. Any correlation scheme for motion estimation, whether implemented in a biological organism or in a machine vision system, does not have the capacity to construct a velocity field estimate with sharp boundaries. For this reason, correlation strategies cannot achieve the type of result desired in this work.

Motion Estimation for Image Compression

Motion estimation has been studied extensively in the context of image compression (Haskell, 1974; Limb and Murphy, 1975a and 1975b; Cafforio and Rocca, 1976; Netravali and Robbins, 1979; Stuller and Netravali, 1979; Stuller, Netravali, and Robbins, 1980; and Jones and Rashid, 1981). Image flow estimation and compression appear to have produced different results because the motivations for the investigations are different, but an equation similar to the image flow equation (Schunck, 1983 and 1984a) has been used to develop motion estimation algorithms for image compression and the motion estimation equations developed for image compression are similar to the iterative equations developed by Horn and Schunck (1981). The apparent difference between the work on image compression and the work on image flow estimation stems from the different motivations for the investigations. Image compression does not require perfect estimation of the motion and does not require the detection of motion boundaries. Any discrepancy between frames caused by inaccurate estimation of the motion is transmitted as a correction. It may be true that a better motion estimation algorithm could reduce the image bandwidth further, but any improvement must be balanced against the added cost of the image compression implementation (Schunck, 1983).

Image Flow Estimation Along Edges

Recent work used edge information in the image flow computation by combining motion information along edge contours (Davis, Wu, and Sun, 1981; Hildreth and Ullman, 1982; and Hildreth, 1984). There are problems with this approach: (1) the velocity vector of an edge is a very powerful clue for grouping edges into contours, but this advantage is dismissed when motion estimation precedes contour grouping, (2) edge fragments can be combined into contours that cross motion boundaries and this will lead to an incorrect motion estimate, and (3) it is not necessary to restrict motion information to edges since it is possible to develop one algorithm that simultaneously estimates the motion at edges and interpolates between edges. The motion estimation algorithm developed by Horn and Schunck (1981) uses a single set of iterative equations that, in effect, automatically balance the computations performed in regions of high and low gradient. When the gradient is large, the information provided by the image flow constraint equation dominates the smoothing process; when the gradient is small, less weight is given to the motion constraint and the iterative equations reduce to the computation of a surface approximation determined by the form of the smoothness constraint. The surface approximation computation iteratively fills in the velocity field using the estimates derived from areas where the gradient is large. Another problem with image flow estimation algorithms that follow edge contours is that they do no better in situations where there are velocity field discontinuities than any of the other algorithms summarized in this section.

2.2.3 Summary of Simple Image Flows

This section presented the standard derivation for the image flow constraint equation, but stressed that the derivation is not suitable for cases where there are image irradiance or velocity field discontinuities. Four examples of image flow estimation algorithms for simple image flows were discussed. None of the algorithms were appropriate for cases where there are image velocity field discontinuities and it is not clear at this point in the presentation whether the image flow equation itself is valid in cases where there are image irradiance or velocity field discontinuities. Extending the equation derivation and algorithm development to such cases will occupy most of the rest of this chapter.

2.3 Discontinuous Image Flow

In this section, the image flow constraint equation for discontinuous image flows will be derived. Since such discontinuities occur in real images, it is important to develop equations and algorithms that can handle the discontinuities.

2.3.1 Surfaces and Projections

Vision involves the projection of surfaces in the scene onto the image plane. The goal of vision is the reconstruction of surface structure and properties from the projections onto the image plane. Image flow estimation involves the reconstruction of the velocity field within the image plane, independent of the motion and structure of the scene surfaces that induce the velocity field; however, the fact that the velocity field is caused by the motion of distinct surfaces in the scene affects the study of image flow. The image flow velocity field may contain discontinuities because there may be separate surfaces in the scene.

Let Σ denote the set of surfaces in the scene and let $(\xi,\eta) \in \Sigma$ denote a point on the surface in the scene. The point, (ξ,η), may or may not be visible. The surfaces are covered with some pattern, $P(\xi,\eta)$. The projection from surface coordinates, (ξ,η), to the image plane, Π, with coordinates, (x,y), is $S: \Sigma \to \Pi$, $(\xi,\eta) \mapsto (x,y)$. The transformation, S, is only defined for the points in Σ that are visible, i.e., projected onto the image plane. For a time-varying image the projection changes with time, $S: \Sigma \times \mathbb{R} \to \Pi$, $(\xi,\eta,t) \mapsto (x,y)$. Now consider the reverse projection from the image plane, Π, to the surface points on Σ that are visible at a given time, t,

$$T: \Pi \times \mathbb{R} \to \Sigma, \quad (x,y,t) \mapsto (\xi,\eta). \tag{10}$$

The discontinuities in the range of this mapping are associated with the jump from one surface, i.e., one contiguous set of ξ,η-coordinates, to another surface in Σ. The position of discontinuities in the image plane changes with time. The image plane pattern, $E(x,y,t)$, is related to the surface pattern, $P(\xi,\eta)$ by

$$E(x,y,t) = P(\xi(x,y,t),\eta(x,y,t)). \tag{11}$$

The purpose of this notation is to make concrete the notion that there are two sources of discontinuity in the image pattern, $E(x,y,t)$. Discontinuities can occur on the surface pattern $P(\xi,\eta)$ or the discontinuity can occur when the mapping, $(x,y,t) \mapsto (\xi,\eta)$, from the visible portion of the scene to the image plane contains discontinuities due to jumps from one surface to another. Neither of these types of discontinuity can be handled by the derivation in Section 2.2.1.

The concept of discontinuities in the mapping from scene surface to image plane coordinates can be generally applied to any intrinsic image. An intrinsic image is the projection of some property of surfaces in the scene such as texture, motion, or surface orientation onto the image plane (Barrow and Tenenbaum, 1978). Surfaces in the scene are separated by depth discontinuities and this structure is projected to the intrinsic images. Any intrinsic image will consist of smoothly varying regions separated by boundaries of step discontinuity. Since the components of the image flow vector field are intrinsic images, each component of the velocity field will consist of smoothly varying regions separated by motion boundaries. This structure is similar to the notion of surface consistency described by Grimson (1983). This intrinsic structure of images motivates the derivation for the image flow equation in the following Section and is a key concept for the algorithm development in Section 2.5 and the smoothing algorithm developed in Section 2.6.

The reader may argue that step changes in image irradiance only introduce technical complications into the derivation of the image flow equation, but the image flow equation will contain δ-functions and it is difficult to work with equations that contain δ-functions as coefficients. Sharp discontinuities do not occur in actual images since the ideal discontinuities are smoothed by the image point spread function of the imaging system or by intentionally smoothing the image to reduce noise, but simple thought experiments containing ideal image irradiance patterns such as step edges can provide key insights to understanding problems in vision and theories of image flow must be able to support such insights. The reformulation of the derivation to handle these technical problems leads to a derivation that addresses the more serious problem of image flow discontinuities caused by occluding surfaces in the scene.

2.3.2 Image Irradiance Discontinuities

A derivation for the image flow equation will be presented that handles discontinuities in the pattern of image irradiance. Consider the situation where a region of the image is moving with velocity, (u,v), and between two points, P_0 and P_1, there is a step change in irradiance. Suppose that the velocity field will cause the patch of image irradiance pattern at point, P_0, to move along some path to point, P_1. The change in irradiance from P_0 to P_1 is given by a simple integral even for a step change in irradiance. If the image irradiance at points, P_0 and P_1, is I_0 and I_1, respectively, then the change in irradiance along the path from P_0 to P_1 is calculated by the line integral

$$I_1 = \int_{P_0}^{P_1} \nabla E \cdot dl + I_0. \qquad (12)$$

The equation is valid even if the gradient must be evaluated across a step change in image irradiance, since any δ-functions will be surrounded by the integral. The change in irradiance over time due to motion past point, P_1, is given by another integral,

$$I_0 = \int_{t_0}^{t_1} \frac{\partial E}{\partial t} \, dt + I_1, \tag{13}$$

where the patch of irradiance that was at P_0 at time, t_0, moves to P_1 at time, t_1. The net change in irradiance, $I_1 - I_0$, must be the same for the two situations,

$$-\int_{t_0}^{t_1} \frac{\partial E}{\partial t} \, dt = \int_{P_0}^{P_1} \nabla E \cdot dl. \tag{14}$$

A thought experiment will illustrate the meaning of Eq.14. Imagine standing on the "surface" of the image irradiance function at point, P_1. With time frozen, walk along the path that the patch of image irradiance at P_0 will follow as it moves from P_0 to P_1 and add up the change in image irradiance. Now imagine standing at P_1, let time proceed and add up the change in image irradiance as the "surface" moves under you. Clearly, the changes accumulated for the two situations must be the same and these changes are equated in Eq.14.

The next step is to transform the integral of change in space in Eq.14 to units of time using speed as the conversion factor. If $(x,y) = (u,v)t$, then

$$-\int_{t_0}^{t_1} \frac{\partial E}{\partial t} \, dt = \int_{t_0}^{t_1} \nabla E \cdot (u,v) dt \tag{15}$$

and the two integrals are over the same domain. Since the argument is true for arbitrary P_0 and P_1 (arbitrary t_0 and t_1 in units of time) the integrands must be equal, so

$$\nabla E \cdot (u,v) + \frac{\partial E}{\partial t} = E_x u + E_y v + E_t = 0, \tag{16}$$

which is the image flow Eq.7 derived in Section 2.2.1.

The essence of the argument is that the change over time at a fixed point must equal the change over space at some fixed time. As long as the change in image intensity is due to the motion displacement, then the change in image intensity detected over time at some point, P, in the image must be the same as the change in image intensity detected by following the path in space that leads to point P. The path along which the line integral is calculated need not be straight. It is only important for the path to be chosen so that it corresponds to the path that a point of image irradiance will follow as it moves across the image plane. The change in space of the image irradiance along the path must be the same as the change in time of the image irradiance that will be seen at the end point of the path as the image irradiance function is moved over the fixed end point of the path. The path corresponds to a stream line from the theory of steady fluid flows.

2.3.3 Velocity Field Discontinuities

The derivation of the previous section can be extended to handle discontinuities in the velocity field due to occlusion. When the image irradiance discontinuity corresponds to a motion boundary, then the derivation is still valid as long as P_0 and P_1 are chosen carefully. The point P_0 can be chosen anywhere on the object, but to avoid the problems caused by occlusion, P_1 must be placed on the opposite side of the motion boundary from P_0, within a small $\varepsilon > 0$ of the boundary. The derivation must avoid invoking conservation of image irradiance along the portion of the path that will be occluded or disoccluded. The result of the derivation is exact in the limit as $\varepsilon \rightarrow 0$.

When the image is sampled in time to form an image sequence, the derivation cannot avoid invoking conservation of image irradiance along the portion of the path affected by occlusion and the image flow equation is incorrect at the motion boundary. This problem will be discussed in

Section 2.4.2. In the continuous space and time domain, the image flow equation is valid at all image points and problems in image flow can be reasoned about in this domain without fear of violating mathematical principles.

2.3.4 Validity of the Image Flow Equation

Only two conditions are required to ensure the validity of the image flow constraint equation: (1) the perceived change in image irradiance at each point in the image plane must be entirely due to translational motion of the image pattern and (2) the image must be smooth except at a finite number of boundary discontinuities. The first condition is satisfied by the image of a translating object that is formed by parallel projection. The image of a translating object formed by perspective projection or the image of a rotating object may not satisfy the first condition, but this is a subtle topic beyond the scope of this presentation. The second condition is a technical assumption required to exclude discontinuities with a structure more complicated than a step change, such as, $\sin(1/x)$ as $x \to 0$. Such functions do not occur as image irradiance distributions, but if such a discontinuity occurred at point, P_1, in the derivation of Section 2.3.3, then the derivation would not be correct even in the limit as $\varepsilon \to 0$ because an infinitesimal displacement of the motion boundary would always occlude an infinite number of sharp changes in the background image irradiance.

2.3.5 Related Work

The psychophysical investigations of Braddick (1974) demonstrate a high standard of performance for a motion segmentation algorithm applied to scenes with motion boundaries. Braddick performed apparent motion experiments with pairs of random-dot images. A central square within one image was displaced to form the second image. The image pairs used by Braddick (1974) differed in element size and displacement distance. Even when the displacement was more than one element, which precludes the obvious dot matching algorithm, the subjects reported that the boundaries of the square were clearly seen. Since no texture or image intensity boundaries were present in the test images, this experiment demonstrates a high level of performance for the short-range motion process. Even though this research is not concerned with discovering the mechanisms of biological vision systems, the results obtained by Braddick influenced this work. Since the short-range motion process in human subjects can detect motion boundaries in scenes where other cues are not present, it must be possible to develop an algorithm for estimating the image flow velocity field while retaining the sharp changes in velocity values that may occur across a motion boundary that corresponds to a switch in surfaces.

A version of the test stimulus used by Braddick (1974) was the key test case for developing an image flow estimation algorithm. A synthetic image was generated with a textured box on a textured background. The foreground box and the background texture were produced by independent sampling from a uniform random number generator in the range, [0,256). The foreground image was overlayed on the background image at a position roughly in the center of the background for the first frame of the test case and at the position one pixel to the right for the second frame of the test case. Uniform noise at various amplitudes was added to these test frames to simulate realistic and greater than realistic noise levels. The additive noise was obtained by independent sampling from a uniform random number generator in the range, $[-N,N]$, where the noise amplitude N was some specified percentage of the maximum pixel amplitude which was 255. After adding a noise sample to a pixel, the pixel was compared with the original pixel range of [0,256). If the pixel was negative, then it was changed to zero; otherwise if the pixel was greater than 255, then it was reduced to 255. This procedure models saturation in black and white. Subsequent work with images obtained from a vidicon camera indicated that a noise level at 5% of the maximum pixel amplitude simulated typical error magnitudes. Since the foreground and background textures had identical statistics, the box could not be seen in either frame alone. Only motion between frames allowed the box to be differentiated from the background. The two frames of this test case are displayed in Figure 1.

2.4 Analysis of Discontinuous Image Flows

Further analysis for image flows with discontinuities will be presented in this section. The image flow equation that was derived in Section 2.2.1 will be transformed in Section 2.4.1 into a polar form that is much easier to use in situations where there are discontinuities in the time-varying image irradiance, $E(x,y,t)$. In Section 2.4.2, the effects of sampling $E(x,y,t)$ to obtain an image sequence will be studied.

2.4.1 Discontinuities in Continuous Image Functions

In this section, the image flow constraint equation will be transformed to polar coordinates with the motion vector represented as speed and direction and the gradient of the image irradiance represented as magnitude and orientation. This representation is convenient for image flows with discontinuities since the polar image flow equation will not contain δ-functions at step discontinuities. Observe that the image flow constraint equation defines a line in velocity space as shown in Figure 2. Since a line is uniquely defined by only two parameters but the image flow equation contains three parameters, there is redundancy. The image flow equation can be formulated so that in situations where there are irradiance discontinuities, the equation contains a ratio of terms that grow individually to infinity while the ratio remains fixed at a single value.

The image flow constraint equation relates the change in space and time of the image irradiance, $E(x,y,t)$, to the velocity vector, (u,v), at a point, (x,y), in the image plane. Referring to Figure 2, the constraint line is uniquely defined by d, the distance of the constraint line from the origin along the line perpendicular to the constraint line, and by the angle, α, with respect to the u-axis. The displacement and angle of the constraint line are

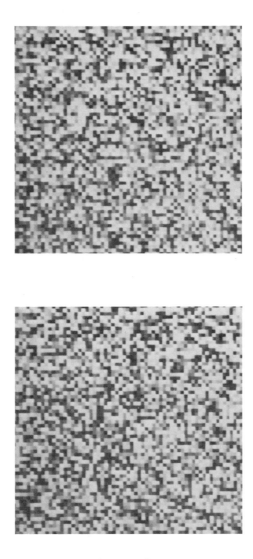

Figure 1. The two frames of the test case used frequently in this work are displayed. The test frames contain a central square that is displaced one pixel to the right between **FRAME-1** and **FRAME-2**. This frame pair does not contain added noise.

given by

$$d = \frac{|E_t|}{\sqrt{E_x^2 + E_y^2}} = \frac{|E_t|}{|\nabla E|};$$

(17)

$$\alpha = \begin{cases} \arctan(E_x, E_y), & \text{if } E_t \geq 0; \\ \arctan(-E_x, -E_y), & \text{if } E_t < 0. \end{cases}$$

The constraint equation in polar coordinates can be derived by noting that the image flow equation contains the dot product of the image irradiance gradient with the velocity vector:

$$E_x u + E_y v + E_t = \nabla E \cdot (u,v) + E_t.$$ (18)

If ρ is the speed of motion, α is the angle of the constraint line, and β is the direction of motion, then the dot product in Eq.18 is

$$\rho |\nabla E| \cos(\alpha - \beta).$$ (19)

Dividing through by the magnitude of the image gradient,

$$d = \rho \cos(\alpha - \beta).$$ (20)

Note that β is constrained to be between $\alpha - \pi/2$ and $\alpha + \pi/2$. Since the displacement, d, of the constraint line from the origin must always be non-negative, the orientation, α, is reflected when $E_t > 0$. The displacement, d,

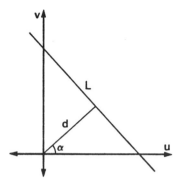

Figure 2. The locus of points satisfied by the simple image flow equation is shown as line, L. The distance d from the origin to L is called the displacement of the constraint line; the angle α is called the orientation of the constraint line.

is the projection of the motion vector onto the line of the gradient of image irradiance and is independent of the magnitude or polarity of the gradient. Imagine a gradual step edge in image irradiance that changes slowly into a step discontinuity. The constraint line displacement, d, is the ratio of two quantities that will tend to infinity as the local change in image irradiance becomes a step discontinuity, but the ratio remains constant. The displacement, d, will remain finite even when the variation in image irradiance becomes a step change. The polar image flow equation remains valid and sensible even when the image is not analytic. Specifically, the image flow equation is valid in the real-world case where the image consists of smooth regions separated by step discontinuities.

There are many conceptual advantages to using the polar form of the image constraint equation: (1) the equation will not contain δ-functions when the image contains step discontinuities; (2) the equation is non-redundant since contrast and opposite gradients are factored out; and (3) the equation makes it easy to study the image flow constraints for various gradient orientations. The partial derivatives of $E(x,y,t)$ with respect to x, y, and t do not explicitly appear in the formula. The partial derivatives enter the formula only as the ratio that defines the constraint line displacement, d. The conceptual obstacle of δ-functions appearing in the formulas during examples with step functions is avoided. The polar representation eliminates redundancy since points that are undergoing the same motion, but have different contrasts or gradients that are reflections of each other will have the same image flow constraint. Since image contrast has been factored from the formula, image flow problems can be studied without adding variations in image contrast to the other effects that must be considered. If the constraint line displacement, d, is treated as a function of α, parameterized by a fixed speed of motion, ρ, and direction of motion, β, then the polar flow equation provides a single function of one variable that shows the magnitude of the projection of the motion vector onto the line of the gradient for any gradient line orientation α. A scatter plot of d versus α, with d and α measurements from image data, provides a dramatic display of the errors in computing the image flow data as will be shown in the next section.

2.4.2 Sampling of Discontinuous Image Flows

The time-varying image, $E(x,y,t)$, is sampled in space and time to produce an image sequence. To preserve the information in the image sequence, the spatial sampling distance must be smaller than the scale of image texture. The temporal sampling period must be much shorter than the scale of time over which the velocity field changes and sufficiently short, relative to the velocity, so that the local displacement is smaller than the local scale of the image texture.

The constraints are sufficient to allow the sampled image sequence to accurately portray the local motion data between motion boundaries, but are not sufficient to capture the motion information at motion boundaries. The sampling errors at motion boundaries can be so severe that no practical sampling constraint can provide useful motion information at a motion boundary. The problem is easily seen in d and α scatter plots for real and synthetic images. Figure 3 is a plot of d versus α for the synthetic test case of a displaced random box on a random background. The image flow parameters, d and α, were computed using the formulas in Eq.17 using partial derivatives that were computed using first differences over a cube in space and time (Horn and Schunck, 1981, Section 7). The foreground box and background have identical statistics: the pixels are generated by a uniform random number generator in the range, [0,256), and are uncorrelated as described in Section 2.3.5. No noise was added to the test images. The points that are offset from the cosine curve are due to the extreme errors in calculating the d and α measurements along the boundary of the displaced box. Some of the error points define an image flow where the minimum speed of motion (the constraint line displacement) is many times the true speed of motion (the height of the cosine curve). Clearly, the problem of computing and using the motion information along a motion boundary is very severe.

The fundamental problem is that the changes in the image gradient along an object boundary that occur over a non-infinitesimal duration of time are inconsistent with the motion that produces the changes. Imagine two successive frames where there is an occluding boundary within the support of the numerical approximation of the gradient. Suppose that between the two frames a significant fraction of the support is occluded. The partial derivative with respect to time will be computed by taking the difference between the two frames averaged over the support of the gradient. Between the two frames the gradient changes drastically. In fact, the change is far more than in the case of a slight shift in the image irradiance function. The change is as drastic as would occur if there were a displacement between frames that was beyond the scale of texture. In effect, occlusion changes the shape of the image irradiance function rather than shifting it within the scale of texture. The problem occurs along the entire boundary of an object, not just along the occluding or disoccluding boundaries. Severe interactions between the image irradiance change along the boundary of an occluding object and the background texture can occur even when the background is not highly textured. For example, the occluding boundary of a moving object could interact with a single step change in the irradiance of the background. The problem would also be caused by time filtering the image sequence with either a discrete or continuous filter or any kind of computation that averages the frames of the image sequence in time.

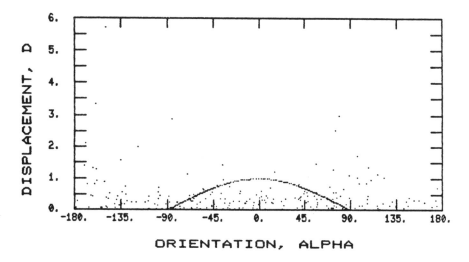

Figure 3. A scatter plot of *d* versus α is shown for the test case of a displaced box without added noise. The points should form a perfect cosine pulse, but there are many points that are distant from the ideal curve. The errors are from the boundary of the displaced box.

Smoothing the images before estimating the partial derivatives and computing the *d* and α data will not solve the problem. Evidence for this is provided in Figure 4. The example of Figure 3 is repeated, except that before the partial derivatives were computed, the test image was smoothed with a rotationally symmetric Gaussian smoothing filter (Canny, 1983). The σ of the smoothing filter was 4.50 pixels. The plot in Figure 4 indicates that in spite of this vigorous amount of smoothing, extreme errors in the image flow equation parameters still occurred along the occluding motion boundary. The errors are so extreme that the images would have to be smoothed until nothing was left but a gray fog in order to smooth away the problem of texture at a motion boundary. The only smoothing operation that could conceivably work would be to remove the occluding object from the image, smooth the background, and replace the object onto a gray, nearly homogeneous background. Of course, such an operation is not possible.

This result indicates that extreme errors in the image flow equation will occur along occluding motion boundaries. Image flow algorithms must be designed to handle these bad measurements. Further details of the problem are presented by Schunck and Horn (1981) and Schunck (1984a).

2.4.3 Directional Selectivity

The relationship between the constraint equation derived above and directional selectivity (Barlow, Hill, and Levick, 1964; Barlow and Levick, 1965; Marr and Ullman, 1981; Batali and Ullman, 1979) will be explained. In directional selectivity, the direction of the gradient of the image irradiance and the sign of the change in image irradiance over time are used to constrain the velocity vector to a half plane. A limit on the maximum velocity further reduces the range of possible velocities to a half circle. The image flow equation constrains the motion vector to a line. In directional selectivity, local constraints are combined by intersecting regions to further constrain the motion to a wedge; with image flow, two intersecting constraint lines restrict the velocity to a single value.

As an example, consider a uniformly shaded rectangle moving across a uniform background of a different shade. The direction of motion is aligned with the long axis of the rectangle. For the purposes of this example, ignore the corners and concentrate on the two pairs of edges that are, respectively, parallel and perpendicular to the direction of motion. The two edges of the rectangle that are perpendicular to the direction of motion, i.e., the leading and trailing edges, each yield the same constraint line, which is a vertical line in velocity space, passing through the actual velocity on the u-axis. The two edges that are parallel to the direction of motion also yield identical constraint lines: the line coincides with the u-axis. The significance of this is that although no change is detected at these edges, the velocity is not constrained to be zero, rather, the velocity vector is constrained to lie along the parallel edges. The constraint line for the two parallel edges intersects the constraint line from the leading and trailing edges at the actual velocity.

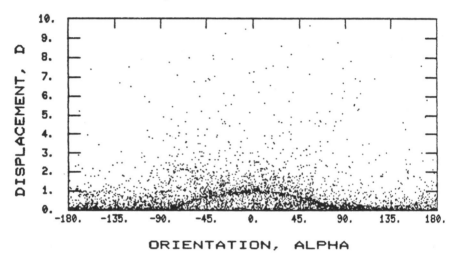

Figure 4. The example of Figure 3 is repeated, except that the test images were smoothed before the partial derivatives for the formulas for d and α were computed. This example shows that smoothing is not sufficient to eliminate the errors in the motion constraint equations along occluding motion boundaries.

Now consider the information provided by directional selectivity. The leading and trailing edges provide identical half circle constraint regions. It is not clear what constraints are provided by the edges parallel to the direction of motion. It is not accurate for the motion detectors at the parallel edges to report no velocity, motion along the edge is possible and it would be misleading to allow the edges parallel to the direction of motion to create half circle shaped constraint regions. In principle, two half circle constraints, back to back, oriented perpendicular to the half

circle constraint generated by the leading and trailing edges could be inter-
sected with the leading and trailing velocity constraint half circle to produce the locus of possible velocities: a line segment extending from the
origin in velocity space along the u-axis to the point where the u-axis intersects the circle of maximum velocity. But edges that are perpendicular to
the direction of motion are not anomalous exceptions and directional selectivity should incorporate the valuable information conveyed by such edges
without resorting to tricks. Directional selectivity ignores the edge contrast
and the magnitude of the temporal change in image irradiance due to
motion. The ratio of change over time to edge contrast is the constraint
line displacement, d, which is the length of the projection of the velocity
vector onto the line of the gradient. The constraint line displacement contains essential information on the possible speeds of motion corresponding
to the possible directions of motion. Directional selectivity ignores the
magnitude of the projection of the motion vector onto the line of the gradient and consequently does not associate a constraint on speed with possible directions of motion. Further information on biological models for
motion detection is provided by Richter and Ullman (1980).

2.4.4 Summary of Discontinuous Image Flows

This section presented the polar form of the image flow equation,
which is a more useful form for working with time-varying images that
may contain image irradiance or velocity field discontinuities. The equation is valid for time-varying images that are continuous in space and time.
When the time-varying image is sampled to produce an image sequence,
the parameters of the image flow equation cannot be correctly computed
along motion boundaries. The errors are severe, cannot be smoothed away,
and the incorrect motion constraints along the boundary must be handled
by an image flow estimation algorithm. The section finished with a comparison of the polar image flow equation with directional selectivity.

2.5 Algorithms for Discontinuous Image Flows

An algorithm called constraint line clustering (Schunck, 1984b) has been developed for estimating the image flow field when the image irradiance pattern or the velocity field may contain discontinuities. Before the algorithm is presented, past work that inspired the development of the algorithm will be discussed.

2.5.1 Background

A function called the convexity function has been proposed for detecting motion boundaries in the image flow field (Nakayama and Loomis, 1974). The algorithm assumed that the image flow field had already been computed. Computing the image flow velocity field in the presence of discontinuities requires that the discontinuities be taken into account which implies at least implicit detection of the discontinuities prior to estimating the velocity field. Batali and Ullman (1979) tried a local consistency check on the motion information provided by directional selectivity to segment random binary images, but the algorithm did not estimate the velocity field. Mutch and Thompson (1984) presented a correspondence algorithm for motion analysis that detected occlusion by looking for significant regions where matches could not be found.

Fennema and Thompson (1979) developed a histogram algorithm for estimating the velocity vector of the most prominent object in the scene by using histogram analysis to combine image flow constraints. The algorithm could only determine the motion of the single most prominent object, could not handle rotation or translation in depth, and assumed that the background was stationary. Since the local structure was lost, the algorithm could not compute an image flow velocity field and this meant that the algorithm was not useful for scenes containing multiple objects. For example, it is not possible to differentiate between (1) the situation where one large object is translating and (2) the situation where two objects, each half the size of the object in situation (1), but separated by some significant distance, are translating. Nevertheless, the algorithm was significant because the histogram technique was immune to the incorrect motion constraints generated along motion boundaries and was not confused by multiple moving objects in the scene as long as one moving object was significantly larger that the others. The Fennema and Thompson algorithm inspired the cluster analysis approach presented in the following sections.

2.5.2 Problem Statement

The constraint line clustering algorithm uses the image flow equation for discontinuous image flows:

$$d = \rho\cos(\alpha - \beta), \tag{21}$$

where $\rho(x,y)$ and $\beta(x,y)$ are the speed and direction of motion, respectively. The velocity vector, (ρ, β), for the motion at any point in the image must lie along the line in velocity space defined by the motion constraint Eq.7 shown in Figure 2. The constraint line is uniquely defined by the displacement, d, of the constraint line from the origin and the orientation, α, of the constraint line. The constraint line displacement has the dimensions of speed and is the minimum speed consistent with the motion constraint.

Assume that the image intensity, $E(x,y,t)$, already incorporates any spatial or temporal filtering performed on the image and has been sampled in space and time to produce a sequence of images, $E(x_i,y_j,t_k)$. The d and α intrinsic image arrays, $d(x_i,y_j)$ and $\alpha(x_i,y_j)$, are computed from the image sequence using the formulas in Eq.17. The partial derivatives are computed by first differences over a cube in space and time (Horn and Schunck, 1981, Section 7) after optionally smoothing with a Gaussian filter. The image flow estimation problem is to compute the intrinsic image arrays $\rho(x_i,y_j)$ and $\beta(x_i,y_j)$ for the speed and direction of motion from the motion measurements in the d and α arrays. Prior knowledge of the

image flow velocity field is not available. By assumption, the image flow
Eq.7 describes the changes in the image sequence so the image flow field
is caused by translational motion and the image is obtained by parallel projection.

2.5.3 Constraint Line Clustering

The constraint line detection algorithm uses a form of cluster analysis
to extract the motion estimate from contradictory data. What could be a
multidimensional cluster analysis problem is transformed into a trivial cluster analysis problem in one dimension. Suppose that for each d and α
measurement, a set of measurements, $\{d_i, \alpha_i\}$, is taken from some spatial
neighborhood of the point that generated d and α. Compute the set of
intersections of each of the neighboring measurements, d_i and α_i, with the
given d and α measurement. All of these intersections lie along the line
defined by d and α. Any constraint lines in $\{d_i, \alpha_i\}$ that are part of the
same region of motion as d and α will tend to intersect the line defined by
d and α in a tight cluster around the true velocity. Any constraint lines in
$\{d_i, \alpha_i\}$ that are from regions of different motion will intersect the line
defined by d and α over a broad range of positions rather than a tight cluster. The bad constraint lines generated along a motion boundary will not
intersect the line defined by d and α at a consistent point. The velocity
estimate for a neighborhood of a given d and α measurement can be computed by one dimensional cluster analysis of intersections along a constraint line.

Computing Constraint Line Intersections

The formula for the position of the intersection along a constraint line is easily derived by first rotating the coordinate system so that the constraint line defined by d and α is parallel to the vertical axis in velocity space. Let the angle between the two constraint lines be denoted by $\phi = \alpha' - \alpha$, as shown in Figure 5. The distance of the intersection along the constraint line defined by d and α is given by

$$b \tan \phi = c. \tag{22}$$

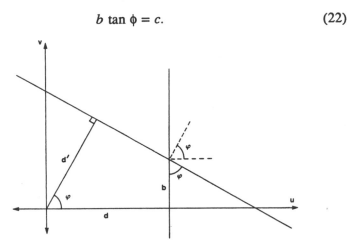

Figure 5. The method for deriving the formula for the position of the intersection of two constraint lines along one of the constraint lines is shown.

The distance c is related to d, d', and ϕ by

$$(d + c)\cos \phi = d', \tag{23}$$

so using Eq. 22 to eliminate c from Eq. 23 yields

$$d \cos \phi + b \sin \phi = d'. \tag{24}$$

Solving for b and substituting the definition for ϕ yields

$$b = \frac{d' - d \cos \phi}{\sin \phi} = \frac{d' - d \cos(\alpha' - \alpha)}{\sin(\alpha' - \alpha)}. \tag{25}$$

This formula provides the position along a constraint line defined by d and α of the intersection of the constraint line with another constraint line defined by d' and α'.

One Dimensional Cluster Analysis

Given a set, $\{b_i\}$, of intersections of the constraint lines within a neighborhood along a given constraint line at the center of the neighborhood, the set of intersections must be analyzed to determine the most consistent subset of intersections that cluster about the likely velocity.

The cluster analysis criterion is motivated and explained by the following example. Imagine that a motion boundary passes almost vertically through a neighborhood just to the left of the center element. At a reasonable pixel resolution, the boundary will most likely pass smoothly through the neighborhood dividing the neighborhood almost in half. The region on the left corresponds to one surface in the scene and the region on the right including the center element corresponds to a different surface. In the extreme case where the boundary passes very close to the center of the neighborhood, almost half of the intersections of neighborhood constraint lines with the constraint line from the center of the neighborhood will be with constraints from the left surface or the motion boundary itself. These intersections will most likely be no where near the velocity of the right surface from which the center constraint line was obtained. Although in the extreme case almost half of the intersections may be useless, at least half of the intersections will correspond to the same region of motion. Since rejecting correct intersections is far less harmful than accepting incorrect intersections, a conservative stand is taken: the algorithm looks for the tightest cluster that contains roughly half of the intersections.

In one dimension, cluster analysis is easy since there is a total ordering of points along a line and cluster analysis reduces to interval analysis. The constraint line clustering algorithm sorts the set of intersections $\{b_i\}$ and then looks for the tightest interval that contains half of the intersections by examining successive pairs of intersections that are about $n/2$ intersections apart, where n is the number of intersections. The algorithm chooses the interval that corresponds to the pair of intersections that are closest together as the estimate of the cluster of intersections that corresponds to the majority velocity in the neighborhood.

The algorithm is very robust and works even if the basic assumption that at least half of the intersections must be from the same region of motion is violated. For example, if the center constraint is from just inside a right angle corner so that only about 1/4 of the intersections are from the same region of motion, the algorithm still works because the intersections from constraints outside the corner are almost uniformly distributed along the constraint line and do not bias the estimate of the tightest cluster (Schunck, 1983).

Estimating the Velocity Vector

The center position, $\hat{b}$, of the tightest cluster along the constraint line is the midpoint of the smallest interval that contains roughly half of the intersections. The estimated speed of motion is given by

$$\hat{v} = \sqrt{d^2 + \hat{b}^2},\tag{26}$$

and the estimated direction of motion is given by

$$\beta = \alpha + \tan(\hat{b}/d).\tag{27}$$

2.5.4 Summary

The constraint line clustering algorithm estimates the velocity vector in successive neighborhoods by intersecting the constraint lines in a neighborhood with the constraint line from the center of the neighborhood and choosing the midpoint of the tightest cluster containing half of the intersections as the motion estimate. A simple rule for reliably selecting the tightest cluster of intersections is used: the tightest cluster is the interval of smallest width along the constraint line that contains half of the intersections. This test is derived from the characteristics of motion boundaries. If the majority of d and α measurements in the neighborhood are from the same side of the boundary as the center d and α measurement, then at least half of the d and α measurements in the neighborhood belong to the same motion as the d and α measurement at the center of the neighborhood. The remaining measurements may be from a different region of motion or from a motion boundary. It is harmless to assume that they are not from the same region of motion, since rejecting a helpful intersection is harmless, but accepting a bad one can ruin the motion estimate. The constraint line clustering algorithm is robust under conditions where there are motion boundaries and high levels of noise. It is an example of ideas in cluster analysis that could be very useful in developing vision algorithms. Schunck (1983) provides details of neighborhood size selection and further discussion of estimation errors.

A version of the test stimulus used by Braddick (1974) was used to test the motion segmentation and estimation algorithm. This test case is called the displaced box example. The results of applying constraint line clustering with 5 by 5, 9 by 9, and 13 by 13 neighborhoods to the displaced box test case with 5% uniform noise is displayed in 6, 7 and 8, respectively. The important point to note is that the estimated velocity field is not improved by increasing the size of the neighborhood in this test case. This means that the quality of the motion estimate is limited by the quality of the d and α measurements, not by the performance of the constraint line clustering algorithm. This is an attractive characteristic from the stand point of this research since it indicates that the algorithm will work well with a given accuracy for the d and α measurements. The subject of measuring the d and α data is within the scope of theories of edge detection and not within the scope of the contribution of this research. The errors introduced by the constraint line clustering algorithm are overshadowed by the errors introduced by the computation of the d and α measurements.

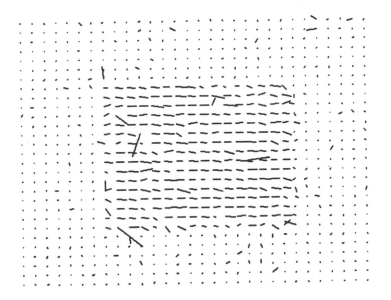

Figure 6. The velocity field after applying constraint line clustering to the test case of the displaced box with 5% uniform noise added is displayed. The neighborhood size was 5 by 5.

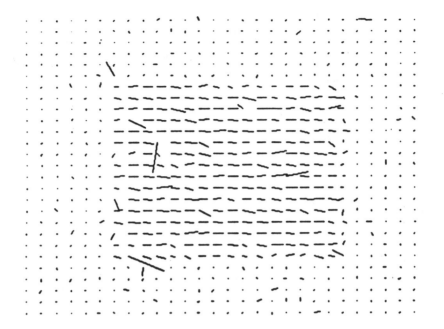

Figure 7. The velocity field after applying constraint line clustering to the test case of the displaced box with 5% uniform noise added is displayed. The neighborhood size was 9 by 9. Note that there is not significant improvement over the field displayed in Figure 6.

2.6 Smoothing Discontinuous Image Flows

An algorithm called surface-based smoothing for improving the velocity field estimate of an image flow vector field is described in this section. The algorithm iteratively computes the motion boundaries in the velocity field and smooths the velocity field estimate between the motion boundaries. A velocity field estimate with sharp motion boundaries can be derived from a noisy initial velocity field estimate. The algorithm is based on a fundamental property of all intrinsic images and can be used to improve the estimate of any intrinsic image.

Surface-based smoothing is able to improve the velocity estimate produced by constraint line clustering because the velocity estimate is good enough to allow the motion boundaries to be detected, so smoothing can be restricted to regions between motion boundaries. The algorithm relies on the fact that the true velocity function is smooth between motion

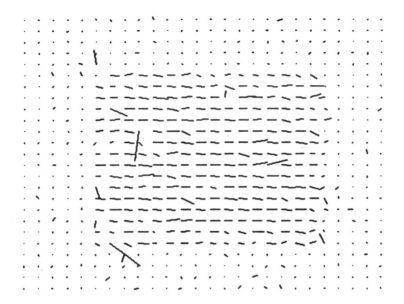

Figure 8. The velocity field after applying constraint line clustering to the test case of the displaced box with 5% uniform noise added is displayed. The neighborhood size was 13 by 13. Note that there is not significant improvement over the field displayed in either Figure 6 or Figure 7. The constraint line clustering algorithm performed very well; the quality of the motion estimate was limited by the d and α measurements.

boundaries because the moving surfaces in the scene that generated the image flow field are smooth between step changes in depth.

2.6.1 Motion Boundary Detection

The velocity field estimated by constraint line clustering is good enough to allow any of the usual edge detection algorithms to be used to detect the motion boundary. Thompson, Mutch and Berzins (1984) presented an algorithm for motion boundary detection that was an extension of the edge detection algorithm of Marr and Hildreth (1980) and presented insights on occlusion analysis. In the work reported in this chapter, the gradient of the Gaussian followed by non-maxima suppression is used to detect boundaries in the components of the image flow field

(Canny, 1983).

After each velocity field component is smoothed with a Gaussian filter, the Prewitt operator is used to compute the x and y partial derivatives of each motion estimate (Nevatia, 1982, p. 106). The Prewitt masks are applied to the smoothed u and v motion data to produce four arrays of boundary data: the x and y partial derivative approximations for the each of the two velocity field components. The partial derivative approximations are used to compute the gradient magnitude and angle for each velocity field component.

Non-maxima suppression reduces the broad ridges and isolated mounds of edge detector output to thin curves and edge fragments. The gradient orientation is reduced to one of four sectors, then the magnitude of the gradient at each point is compared with the magnitude of its two neighbors along the line of the sector. If either of the neighbors of the center element have larger magnitudes, then the magnitude of the center element is set to zero. The total number of edge detector output values that are above some threshold is decreased. This makes more of the velocity field data available to the smoothing algorithm. Non-maxima suppression also produces paths through weak boundaries that allow the velocity field estimates on opposite sides of the boundaries to mix, because weak boundaries can have erratic gradient directions and the edge fragments can cancel one another.

2.6.2 Velocity Field Smoothing

The smoothing problem is formulated as a minimization problem with two components: (1) a cost for the lack of smoothness in the velocity field between motion boundaries, and (2) a cost on the deviation of the smoothed velocity field from the velocity field estimated by constraint line clustering. This approach was motivated by the work of Horn and Schunck (1981) and Grimson (1981a, 1981b, and 1981c). Other methods for implementing the smoothing equations are discussed by Terzopoulos

(1982) and a discussion of viewpoint invariant surface interpolation is provided by Blake (1984). The mathematics of smoothing operators is presented by Brady and Horn (1981). Surface interpolation at multiple scales of resolution is discussed by Terzopoulos (1983).

Let u and v denote the x and y components of the velocity field estimate obtained by constraint line clustering, let $\bar{u}$ and $\bar{v}$ denote the smoothed velocity field estimate. The cost assigned to the deviation of the smoothed velocity field estimate from the original velocity field estimate is

$$(u - \bar{u})^2 + (v - \bar{v})^2. \tag{28}$$

The cost assigned to lack of smoothness is the sum of the squares of the magnitudes of the gradients of the x and y velocity field components,

$$\bar{u}_x^2 + \bar{u}_y^2 + \bar{v}_x^2 + \bar{v}_y^2, \tag{29}$$

as used by Horn and Schunck (1981). The combined optimization criterion is

$$\int\int \kappa^2(\bar{u}_x^2 + \bar{u}_y^2 + \bar{v}_x^2 + \bar{v}_y^2) + (u - \bar{u})^2 + (v - \bar{v})^2 \, dx \, dy, \tag{30}$$

where κ^2 weights the degree of smoothing versus the degree of deviation from the original velocity field estimate. This optimization problem is easily solved by the calculus of variations (Courant and Hilbert, 1937, pp. 191-192). The result is a pair of linear partial differential equations:

$$\kappa^2 \nabla^2 \overline{u} - u = 0$$

and

$$\kappa^2 \nabla^2 \overline{v} - v = 0.$$

Unlike the iterative equations of Horn and Schunck (1981, Section 9), these equations are linear and decoupled. The equations can be solved iteratively by approximating the Laplacian with a nine point mask (Ames, 1977, p. 128),

$$\nabla^2 \approx \begin{bmatrix} 1 & 4 & 1 \\ 4 & -20 & 4 \\ 1 & 4 & 1 \end{bmatrix}. \tag{32}$$

Using this approximation, the iterative equations are easily obtained:

$$\overline{u}^{n+1} = \frac{1}{1 + 20\kappa^2} [u^n + \kappa^2 S(\overline{u}^n)]$$

and

$$\overline{v}^{n+1} = \frac{1}{1 + 20\kappa^2} [v^n + \kappa^2 S(\overline{v}^n)], \tag{33}$$

where the function $S(\cdot)$ represents the computation of the eight point surround of the approximation to the Laplacian.

These smoothing equations should only be used over the interior of a region of slowly varying velocity. The equations should never by applied across a motion boundary since to do so would blur the boundaries by combining motion estimates across motion boundaries. Natural boundary conditions must be used when the support of the approximation to the Laplacian is next to a motion boundary. Since the shape of the motion boundary is arbitrary, it is very difficult to develop and apply the approximations to the Laplacian that should be used at a motion boundary. A simple scheme was used to approximate natural boundary conditions near a motion boundary: the velocity field value at the center of the Laplacian is used in place of the velocity field value at any point in the support of the Laplacian that is a boundary point. This scheme is simple to implement and can be applied to boundaries of arbitrary shape.

2.6.3 Interleaved Detection and Smoothing

The algorithm for detecting motion boundaries and the algorithm for smoothing the velocity field estimate between motion boundaries are interleaved so that each computation can benefit from the improved information provided by the other. The computation begins with the original velocity field estimate computed by constraint line clustering. The x and y components of the velocity field are contained in separate arrays. The motion edge detector is applied to each velocity component to obtain separate arrays of edge information for each velocity component. One iteration of the smoothing operator is applied in place to each component of the velocity field, using the boundary information for that component, to obtain the new smoothed version of each velocity field component. The border of the arrays is filled using data from the rows and columns just inside the border to approximate natural boundary conditions along the perimeter of the field of view. Subsequent iterations begin by applying the motion boundary detection operator to the smoothed velocity field estimate. The smoothing that was performed between motion boundaries over the interior of surfaces will have reduced the variation in the velocity field. Since the variation is less, the output of the boundary detection operator in these regions will be weaker. Near a motion boundary, the local variation in the velocity field will be more apparent: the velocity values on either side of the boundary will be more consistently the same and the velocity values on opposite sides of the boundary will be more consistently different. The velocity values in the interior of the surfaces corresponding to objects and the background will not wander far from the original values estimated by constraint line clustering since there is a cost on the deviation of the smoothed velocity estimates from the original velocity estimates. This anchors the velocity estimates to the original motion data. As successive iterations are performed, the height of the boundary operator output will increase along a motion boundary, decrease over the interior of objects and background points between boundaries, and the smoothing of the velocity estimate will be confined to a well within which a smooth velocity field estimate will be achieved independent of the velocity field values at points on the other side of the motion boundary. Examples and details of the algorithm development are provided by Schunck (1983 and 1984c). Related work on relaxation algorithms has been presented by Geman and Geman (1984).

The boundary estimation and surface smoothing computation was applied to the test case shown in Figure 6. The results of surface-based smoothing are displayed in Figures 9 through 11. The results show that the interleaved iterations of boundary estimation and surface smoothing can improve the velocity field estimate and produce a clean estimate of the motion boundary.

2.7 Summary and Conclusions

This chapter has presented work on image flow estimation within a framework that differentiates between image flows that may be subject to discontinuities in the image irradiance or velocity field and image flows that are restricted from exhibiting such phenomena. For simple image flows that do not contain discontinuities of either kind, the image flow equation can be derived by Taylor series expansion and image flow estimation algorithms have been developed through regularization methods (Poggio, 1985). Although the development of algorithms for simple image flows has produced valuable milestones for vision research, image flows from real scenes will contain discontinuities and it is not clear how

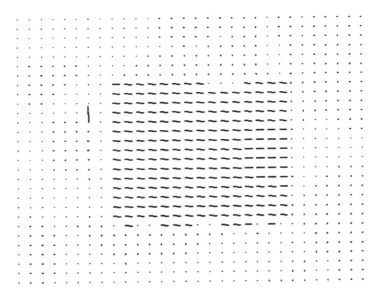

Figure 9. The velocity field computed after 16 smoothing iterations have been performed on the velocity field shown in Figure 9 is displayed. The motion edges were computed using the Prewitt x and y partial derivative approximations applied to the motion components after smoothing with an 8 by 8 Gaussian filter. The motion edge threshold was 1000 and 16.0 was used for the smoothing weight.

Figure 10. The plot is the boundary of the u component of the velocity field shown in Figure 9. Note that there are no false boundary points.

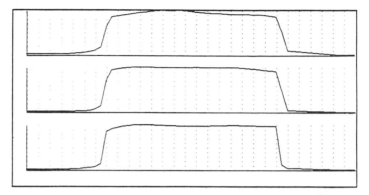

Figure 11. Three horizontal slices through the *u* component of the smoothed velocity field from Figure 9 are graphed. The deflection of the estimated value from the true value is no greater than 3%

algorithms developed for simple image flows can be extended to the more realistic cases. A derivation of the image flow equation for discontinuous image flows was presented and the equation was reformulated in polar coordinates. The polar form is more appropriate for discontinuous image flows. A one-dimensional cluster analysis algorithm for the estimation of possibly discontinuous image flows was presented and the chapter concluded with a presentation of a smoothing algorithm that improves velocity field estimates without blurring the motion boundaries.

Acknowledgements

This report describes research done at the Artificial Intelligence Laboratory of the Massachusetts Institute of Technology. Support for this research was provided in part by the Advanced Research Projects Agency under Office of Naval Research contract N00014-80-C-0505. This chapter is based on a thesis submitted in partial fulfillment of the requirements of the degree of Doctor of Science in the Department of Electrical Engeneering and Computer Science at the Massachusetts Institute of Technology in May 1983.

The manuscript for this chapter was prepared while the author was at General Motors Research Laboratories.

References

Ames, W.F., (1977) **Numerical Methods for Partial Differential Equations**, 2nd ed., New York: Academic Press.

Barlow, H.B., Hill, R.M., and Levick, W.R., (1964) 'Retinal ganglion cells responding selectively to direction and speed of image motion in the rabbit,' *J. of Physiology*, vol. 173, pp. 377-407.

Barlow, H.B. and Levick, W.R., (1965) 'The mechanism of directionally selective units in rabbit's retina,' *J. of Physiology*, vol. 178, pp. 477-504.

Barrow, H.G. and Tenenbaum, J.M., (1978) 'Recovering intrinsic scene characteristics from images,' in **Computer Vision Systems**, A.R. Hanson and E.M. Riseman (eds.), Academic Press, New York.

Batali, J. and Ullman, S., (1979) 'Motion detection and analysis,' in **Image Understanding**, L.S. Baumann (ed.), Science Applications, Arlington, Virginia, pp. 69-75,

Bers, K.H., Bohner, M., and Gerlach, H., (1980) 'Object detection in image sequences,' *Proc. Int. Joint Conf. Pattern Recognition*, pp. 1317-1319.

Blake, A., (1984) 'Reconstructing a visible surface,' *Proc. Nat. Conf. Artificial Intelligence*, pp. 23-26.

Braddick, O., (1974) 'A short-range process in apparent motion,' *Vision Research*, vol. 14, pp. 519-527.

Brady, M., and Horn, B.K.P., (1981) 'Rotationally symmetric operators for surface interpolation,' Memo 654, Artificial Intelligence Laboratory, M.I.T.

Buckner, E., (1976) 'Elementary movement detectors in an insect visual system,' *Biological Cybernetics,* vol. 24, pp. 85-101.

Buxton, B.F., Buxton, H., Murray, D.W., and Williams, N.S., (1983) '3D solutions to the aperture problem,' Tech. Rep., GEC Research Laboratories.

Cafforio, C., and Rocca, F., (1976) 'Methods for measuring small displacements of television images,' *IEEE Trans. on Information Theory,* vol. IT-22, pp. 573-579.

Canny, J.F., (1983) 'Finding edges and lines in images,' Tech. Rep. 720, Artificial Intelligence Laboratory, M.I.T.

Clocksin, W.F., (1978) 'Determining the orientation of surfaces from optical flow,' *Proc. 3rd AISB Conf.,* pp. 93-102.

Clocksin, W.F., (1980) 'Perception of surface slant and edge labels from optical flow: A computational approach.' *Perception,* vol. 9, pp. 253-269.

Courant, R. and Hilbert, D., (1937) **Methods of Mathematical Physics,** vol. 1, Wiley-Interscience, New York.

Davis, L.S., Wu, Z., and Sun, H., (1981) 'Contour-based motion estimation,' Tech. Rep., Computer Vision Laboratory, University of Maryland.

Fennema, C.L., and Thompson, W.B., (1979) 'Velocity determination in scenes containing several moving objects,' Computer Graphics and Image Proccessinf, vol. 9, pp. 301-315.

Flinchbaugh, B.E., and Chandrasekaran, B., (1981) 'A theory of spatio-temporal aggregation for vision,' *Artificial Intelligence,* vol. 17, pp.

387-407. reprinted in **Computer Vision, M.** Brady (ed.), North-Holland, Amsterdam.

Geman, S., and Geman, D., (1984) 'Stochastic relaxation, Gibbs distributions, and the Bayesian restoration of images,' *IEEE Trans. on Pattern Analysis and Machine Intelligence,* vol. PAMI-6, pp. 721-741.

Grimson, W.E.L., (1981a) **From Images to Surfaces: A Computational Study of the Human Early Vision System,** M.I.T. Press, Cambridge.

Grimson, W.E.L., (1981b) 'A computational theory of visual surface interpolation,' Memo 613, Artificial Intelligence Laboratory, M.I.T.

Grimson, W.E.L., (1981c) 'The implicit constraints of the primal sketch,' Memo 663, Artificial Intelligence Laboratory, M.I.T.

Grimson, W.E.L., (1983) 'Surface consistency constraints in vision,' *Computer Vision, Graphics and Image Processing,* vol. 24, pp. 28-51.

Haskell, (1974) 'Frame-to-frame coding of television pictures using two-dimensional Fourier transforms,' *IEEE Trans. on Information Theory,* vol. IT-20, pp. 119-120.

Hildreth, E.C., (1984) 'The computation of the velocity field,' *Proc. Royal Society London,* B, vol. 221, pp. 189-220.

Hildreth, E.C., and Ullman, S., (1982) 'The measurement of visual motion,' Memo 699, Artificial Intelligence Laboratory, M.I.T.

Hirzinger, G., Landzettel, K., and Snyder, W., (1980) 'Automated TV tracking of moving objects: The DFVLR tracker and related approaches,' *Proc. Int. Joint Conf. Pattern Recognition,* pp.1255-1261.

Hoffman, D.D., (1980) 'Inferring shape from motion fields,' Memo 592,

Artificial Intelligence Laboratory, M.I.T.

Horn, B.K.P., and Bruss, A.R., (1983) 'Passive navigation,' *Computer Vision, Graphics and Image Processing*, vol. 21, pp. 3-20.

Horn, B.K.P., and Schunck, B.G., (1981) 'Determining optical flow,' *Artificial Intelligence*, vol. 17, reprinted in **Computer Vision**, M. Brady (ed.), North-Holland, Amsterdam, pp. 185-203.

Jain, R., Martin, W.N., and Aggarwal, J.K., (1979) 'Segmentation through the detection of changes due to motion,' *Computer Graphics and Image Processing*, vol. 11, pp. 13-34.

Jain, R., Militzer, D., and Nagel, H.-H., (1977) 'Separating non-stationary from stationary scene components in a sequence of real world TV-images,' *Proc, Int. Joint Conf. Artificial Intelligence*, pp. 612-618.

Jain, R., and Nagel, H.-H., (1979) 'On the analysis of accumulative difference pictures from image sequences of real world scenes,' *IEEE Trans. on Pattern Analalysis and Machine Intelligence*, vol. PAMI-1, pp. 206-214.

Jain, R., (1983) 'Difference and accumulative difference pictures in dynamic scene analysis,' Rep. GMR-4548, General Motors Research Laboratories.

Jones, R.A., and Rashid, H., (1981) 'Residual recursive displacement estimation,' *Proc Conf. Pattern Recognition and Image Processing*, pp. 508-509.

Koenderink, J.J., and van Doorn, A.J., (1975) 'Invariant properties of the motion parallax field due to the movement of rigid bodies relative to an observer,' *Optica Acta*, vol. 22, pp. 773-791.

Koenderink, J.J., and van Doorn, A.J., (1976) 'Local structure of movement parallax of the plane,' *J. Optical Society America,* vol. 66, pp. 717-723.

Limb, J.O., and Murphy, J.A., (1975a) 'Measuring the speed of moving objects from television signals,' *IEEE Trans. on Communication,* vol. COM-23, pp. 474-478.

Limb, J.O., and Murphy, J.A., (1975b) 'Estimating the velocity of moving images in television signals,' *Computer Graphics and Image Processing,* vol. 4, pp. 311-327.

Longuet-Higgins, H.C., and Prazdny, K., (1980) 'The interpretation of moving retinal image,' *Proc. Royal Society London,* vol. B 208, pp. 385-397.

Marr, D., and Hildreth, E., (1980) 'Theory of edge detection,' *Proc. Royal Society London,* vol. B 207, pp. 187-217.

Marr, D., and Ullman, S., (1981) 'Directional selectivity and its use in early visual processing,' *Proc. Royal Society London,* vol. B 211, pp. 151-180.

Mutch, K.M., and Thompson, W.B., (1984) 'Analysis of accretion and deletion at boundaries in dynamic scenes,' Tech. Rep. 84-7, Department of Computer Science, University of Minnesota, Minneapolis.

Nagel, H.-H., (1980) 'From digital picture processing to image analysis,' Proc. Int. Conf. Image Analysis and Processing.

Nakayama, K., and Loomis, J., (1974) 'Optical velocity patterns, velocity-sensitive neurons, and space perception: A hypothesis,' *Perception,* vol. 3, pp. 63-80.

Netravali, A.N., and Robbins, J.D., (1979) 'Motion-compensated television coding: Part I,' *Bell System Tech. J.*, vol. 58, pp. 631-670.

Nevatia, R., (1982) **Machine Perception**, Prentice-Hall, Englewood Cliffs, NJ.

Poggio, T., and Reichardt, W., (1976) 'Visual control of orientation behavior in the fly,' *Quart. Rev. of Biophysics*, vol. 9, pp. 377-438.

Poggio, T., (1985) 'Early vision: From computational structure to algorithms and parallel hardware,' *Computer Vision, Graphics and Image Processing*, vol. 31, pp. 139-155.

Prazdny, K., (1980) 'Egomotion and relative depth map from optical flow,' *Biological Cybernetics*, vol. 36, pp. 87-102.

Price, C., Snyder, W., and Rajala, S., (1981) 'Computer tracking of moving objects using a Fourier-domain filter based on a model of the human visual system,' *Proc, Conf. Pattern Recognition and Image Processing*, pp. 98-102.

Reichardt, W., and Poggio, T., (1979) 'Figure-ground discrimination by relative movement in the visual system of the fly,' *Biological Cybernetics*, vol. 35, pp. 81-100.

Richter, J., and Ullman, S., (1980) 'A model for the spatio-temporal organization of X and Y-type ganglion cells in the primate retina,' Memo 573, Artificial Intelligence Laboratory, M.I.T.

Roach, J.W. and Aggarwal, J.K., (1979) 'Computer tracking of objects moving in space.' *IEEE Trans. on Pattern Analalysis and Machine Intelligence*, vol. PAMI-1, pp. 127-135.

Schalkoff, R.J., and McVey, E.S., (1982) 'A model and tracking algorithm for a class of video targets,' *IEEE Trans. on Pattern Analalysis and Machine Intelligence,* vol. PAMI-4, pp. 2-10.

Schetzen, M., (1980) **The Voltera and Wiener Theories of Nonlinear Systems,** John Wiley & Sons, New York.

Schunck, B.G., (1983) 'Motion segmentation and estimation,' Department of Electrical Engineering and Computer Science, M.I.T., PhD. Dissertation.

Schunck, B.G., (1984a) 'The motion constraint equation for optical flow,' *Proc. Int. Joint Conf. on Pattern Recognition,* Montreal, pp. 20-22.

Schunck, B.G., (1984b) 'Motion segmentation and estimation by constraint line clustering,' *Proc. 2nd IEEE Workshop on Computer Vision,* Annapolis, Maryland, pp. 58-62.

Schunck, B.G., (1984c) 'Surface-based smoothing of optical flow fields,' *Proc. Conf. on Intelligent Systems and Machines,* Oakland University, Rochester, Michigan, pp. 107-111.

Schunck, B.G., and Horn, B.K.P., (1981) 'Constraints on optical flow computation,' *Proc. Conf. Pattern Recognition and Image Processing,* pp. 205-210.

Snyder, W.E., Rajala, S.A., and Hirzinger, G., (1980) 'Image modelling: The continuity assumption and tracking,' *Proc. Int. Joint Conf. Pattern Recognition,* pp. 1111-1114.

Stuller, J.A., and Netravali, A.N., (1979) 'Transform domain motion estimation,' *Bell System Tech. J.,* vol. 58, pp. 1673-1702.

Stuller, J.A., Netravali, A.N., and Robbins, J.D., (1980) 'Interframe television coding using gain and displacement compensation,' *Bell System*

Tech. J., vol. 59, pp. 1227-1240.

Terzopoulos, D., (1982) 'Multilevel reconstruction of visual surfaces: Variational principles and finite element representations,' Memo 671, Artificial Intelligence Laboratory, M.I.T.

Terzopoulos, D., (1983) 'Multilevel computational processes for visual surface reconstruction,' *Computer Vision, Graphics and Image Processing*, vol. 24, pp. 52-96.

Thompson, W.B., Mutch, K.M., and Berzins, V.A., (1984) 'Dynamic occlusion analysis in optical flow fields,' Tech. Rep. 84-6, Department of Computer Science, University of Minnesota.

Tsai, R.Y., and Huang, T.S., (1982) 'Uniqueness and estimation of three-dimensional motion parameters of rigid objects with curved surfaces,' *Proc. Conf. Pattern Recognition and Image Processing*, pp. 112-118.

Tsai, R.Y., Huang, T.S., and Zhu, W.-L., (1982) 'Estimating three-dimensional motion parameters of a rigid planar patch, II: Singular value decomposition,' *IEEE Trans. on Acoustics, Speech, and Signal Processing*, vol. ASSP-30, pp. 525-534.

Ullman, S., (1979a) 'The interpretation of structure from motion,' *Proc. Royal Society London*, vol. B 203, pp. 405-426.

Ullman, S., (1979b) **The Interpretation of Visual Motion**, M.I.T. Press, Cambridge.

Ullman, S., (1979c) 'Relaxation and constrained optimization by local processes,' *Computer Graphics and Image Processing*, vol. 10, pp. 115-125.

Waxman, A., and Ullman, S., (1983) 'Surface structure and 3-d motion from image flow: A kinematic analysis,' Tech. Rep., Center for Automation Research, University of Maryland.

Waxman, A.M. and Sinha, S.S., (1984) 'Dynamic stereo: Passive ranging to moving objects from relative image flows,' Tech. Rep., Computer Vision Laboratory, University of Maryland.

Waxman, A.M., (1984) 'An image flow paradigm,' Tech. Rep., Computer Computer Vision Laboratory, University of Maryland.

Waxman, A.M., and Wohn, K., (1984) 'Contour evolution, neighborhood deformation and global image flow: Planar surfaces in motion,' Computer Vision Laboratory, University of Maryland.

Yalamanchili, S., Martin, W.N., and Aggarwal, J.K., (1982) 'Extraction of moving object descriptions via differencing,' *Computer Vision, Graphics and Image Processing,* vol. 18, pp. 188-201.

CHAPTER 3

A Computational Approach to the
Fusion of Stereopsis and Kineopsis

Amar Mitiche

3.1 Introduction

Vision research in fields as diverse as computer science, psychology, and neurophysiology, has led to the emergence of *stereopsis* and *kineopsis* as the two principal views which explain some of the mechanisms of space perception.

Stereopsis, which is primarily supported by the work of Julesz (1971) and some neurophysiological evidence (Barlow, Blakemore and Pettigrew, 1967), refers to the process of recovering depth from the retinal disparity field. In this context, disparity is the difference in the positions of the images of a point in space resulting from the difference in stereoscopic perspectives. Computational processes that use stereoscopic imaging must therefore solve the correspondence problem (Aggarwal, Davis and Martin, 1982; Marr and Poggio, 1979). With a particular location in one of the stereoscopic images, the correspondence process must associate the location in the other image which represents the projection of the same element in space. Such locations are generally characteristic in the images as textured areas, edges, segments, or corners. Once correspondence is achieved, depth is recovered quite simply (Duda and Hart, 1973).

Kineopsis (Nakayama and Loomis, 1974) refers to the process which operates on the optical flow pattern for space perception. Optical flow is the field of apparent velocities of images of object points in space, i.e., optical velocities, due to relative motion of viewing system and objects. Optical flow is the focus of a number of studies and there is a significant interest in developing methods to compute it reliably (Fennema and Thompson, 1979; Horn and Schunck, 1981; Paquin and Dubois, 1983; Barnard and Thompson, 1980; Nagel, 1982; Nagel, 1983; Wohn, Davis, and Thrift, 1983). There is also interest in establishing computational models of kineopsis (Bruss and Horn, 1983; Lawton and Rieger, 1983; Longuet-Higgins and Prazdny, 1980; Prazdny, 1983; Waxman, 1983; Gibson, Gibson, Smith, and Flock, 1959; Thompson and Barnard, 1981; Thompson, 1980; Potter, 1972; Ballard and Kimball, 1983; Mitiche, 1984b). Methods have been proposed where optical flow is the basis for (a) recovering relative depth and motion of objects in space, (b) segmenting scenes into moving and stationary objects, and (c) predicting position of features in images. Computational kineopsis is confronted with two major problems: (a) determination of structure and motion in space from image positions and optical velocities, and (b) reliable computation of optical velocities.

Stereopsis and kineopsis are generally believed to control different aspects of space perception. It has been suggested that retinal disparity controls perception of the near environment whereas optical flow controls the more distant environment (Nakayama and Loomis, 1974). Some authors have suggested that stereopsis and kineopsis operate in similar fashion regardless of what they actually control. Gibson (Gibson, Gibson, Smith, and Flock; 1959), for instance, considered stereoscopic disparity to be a special case of motion disparity, contending that the only difference is that the former is simultaneous and the latter successive in nature.

Perhaps because of such atomistic views of space perception, stereopsis and kineopsis have been traditionally studied separately and the problem of combining their mechanisms or effects has been ignored. An important question we ask here is: what can we gain from combining stereopsis and kineopsis that cannot be gained, or would be difficult to gain, from either stereopsis or kineopsis considered independently? We propose an answer by demonstrating the following:

(1) Optical flow can be integrated with the stereopsis mechanism to determine three-dimensional motion. To demonstrate this effect, we will develop an analytical expression of three-dimensional motion in terms of depth, optical flow, and stereoscopy parameters.

(2) Once three-dimensional motion is determined, rigid motion and structure can be characterized and computed directly by a local test based on optical flow and depth. In some circumstances, this computation cannot be achieved using optical flow alone or depth alone.

Our approach, which could be called *stereokineopsis*, therefore provides an integrated computational process for the perception of space.

The remainder of this chapter is organized as follows: Section 3.2 describes how three-dimensional motion can be obtained from depth, optical flow, and stereoscopy parameters. Section 3.3 treats the problem of determining rigidity from motion, and Section 3.4 gives examples. Finally, Section 3.5 contains a summary.

3.2 Integrating Optical Flow to Stereopsis for Motion

We will adopt the configuration of cameras illustrated in Figure 1. Each camera is represented by a central projection model that does not invert images. We will refer to the cameras as the *first* and the *second* camera. The first and second camera have image planes, I_1 and I_2, and projection centers, O_1 and O_2, respectively. The world-coordinate system is $S=(O,x,y,z)$. The first and second camera coordinate systems are $S_1=(O_1,X_1,Y_1,Z_1)$ and $S_2=(O_2,X_2,Y_2,Z_2)$, while (c_1,x_1,y_1) and (c_2,x_2,y_2) denote the I_1- and I_2-coordinate systems, respectively. The focal lengths of the first and second camera are F_1 and F_2, respectively. The optical axes of the cameras are aligned with the Z_1-axis and the Z_2-axis. In the following it is

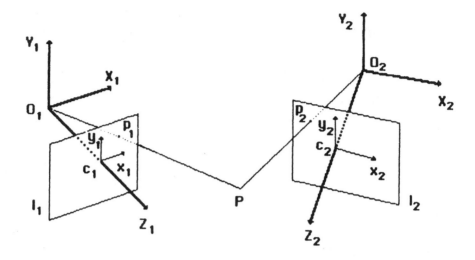

Figure 1. Geometric configuration.

convenient to assume, without loss of generality, that reference systems S and S_1 coincide and that $F_1=F_2=1$.

Let P be a point in space, with coordinates (X_1,Y_1,Z_1) in S_1 and (X_2,Y_2,Z_2) in S_2. Let the image of P be p_1 on I_1 with coordinates (x_1,y_1) and be p_2 on I_2 with coordinates (x_2,y_2). Finally, let the transform taking S_1 onto S_2 be decomposed into a rotation, R, about an axis through the origin, O_1, and a translation, T. The rotation is described by the matrix:

$$R = \begin{bmatrix} a_{11} & a_{12} & a_{13} \\ a_{21} & a_{22} & a_{23} \\ a_{31} & a_{32} & a_{33} \end{bmatrix},$$

and the translation by the vector, $T = (X_T,Y_T,Z_T)$. With these notations we can write

$$(X_2, Y_2, Z_2) = (X_1, Y_1, Z_1) R + T,$$

or, in expanded form,

$$X_2 = a_{11}X_1 + a_{21}Y_1 + a_{31}Z_1 + X_T,$$

$$Y_2 = a_{12}X_1 + a_{22}Y_1 + a_{32}Z_1 + Y_T, \quad \text{and}$$

$$Z_2 = a_{13}X_1 + a_{23}Y_1 + a_{33}Z_1 + Z_T .$$

Since

$$x_2 = \frac{X_2}{Z_2} \quad \text{and} \quad y_2 = \frac{Y_2}{Z_2}$$

we have

$$x_2 = \frac{a_{11}X_1 + a_{21}Y_1 + a_{31}Z_1 + X_T}{a_{13}X_1 + a_{23}Y_1 + a_{33}Z_1 + Z_T} .$$

Dividing the numerator and denominator of the right-hand side by Z_1, we obtain

$$x_2 = \frac{a_{11}x_1 + a_{21}y_1 + a_{31} + \dfrac{X_T}{Z_1}}{a_{13}x_1 + a_{23}y_1 + a_{33} + \dfrac{Z_T}{Z_1}} ,$$

similarly,

$$y_2 = \frac{a_{12}x_1 + a_{22}y_1 + a_{32} + \dfrac{Y_T}{Z_1}}{a_{13}x_1 + a_{23}y_1 + a_{33} + \dfrac{Z_T}{Z_1}} \; .$$

We can therefore write x_2 and y_2 as functions of x_1, y_1 and Z_1 as follows:

$$x_2 = \alpha \, (x_1, y_1, Z_1), \quad \text{and}$$

$$y_2 = \beta \, (x_1, y_1, Z_1).$$

Thus, knowing x_1 and y_1 we can determine x_2 and y_2 from depth and camera calibration parameters. The above notation can now be augmented to incorporate a dependence on time as follows:

$$x_2(t) = \alpha(x_1(t), y_1(t), Z_1(t)), \quad \text{and}$$

$$y_2(t) = \beta(x_1(t), y_1(t), Z_1(t)).$$

But because optical flow at p_2 is (u_2, v_2) with

$$u_2 = \frac{dx_2}{dt} \quad \text{and} \quad v_2 = \frac{dy_2}{dt} \; ,$$

and we can write

$$u_2 = \frac{\partial \alpha}{\partial x_1} \cdot \frac{dx_1}{dt} + \frac{\partial \alpha}{\partial y_1} \cdot \frac{dy_1}{dt} + \frac{\partial \alpha}{\partial Z_1} \cdot \frac{dZ_1}{dt}$$

$$= \frac{\partial \alpha}{\partial x_1} u_1 + \frac{\partial \alpha}{\partial y_1} v_1 + \frac{\partial \alpha}{\partial Z_1} \cdot \frac{dZ_1}{dt} , \tag{1}$$

where (u_1, v_1) is the optical flow at p_1, ∂ designates the partial derivative and d the ordinary derivative. Similarly for v_2 we have

$$v_2 = \frac{\partial \beta}{\partial x_1} u_1 + \frac{\partial \beta}{\partial y_1} v_1 + \frac{\partial \beta}{\partial Z_1} \cdot \frac{dZ_1}{dt}. \tag{2}$$

We see that if points, p_1 in I_1 and p_2 in I_2, are in correspondence, then their optical velocities are related by the above equations. If this correspondence is established and stereoscopy parameters are known then the functions, $\alpha(t)$ and $\beta(t)$, and their derivatives appearing in Eqs. 1 and 2 are completely specified. We can then determine $Z'_1 = \dfrac{dZ_1}{dt}$ from either Eq. 1, which involves the horizontal components of u_1 and u_2, and the vertical component, v_1, at image points p_1 and p_2, or Eq. 2, which involves the vertical components of optical flow, v_1 and v_2, and the horizontal component, u_1.

Once Z'_1 is computed, we can obtain the other two components of three-dimensional velocity as follows. Since

$$X_1 = x_1 Z_1, \quad \text{and}$$
$$Y_1 = y_1 Z_1$$

then

$$X'_1 = x_1 Z'_1 + u_1 Z_1,$$
$$Y'_1 = y_1 Z'_1 + v_1 Z_1.$$

To justify this development, consider that if S_1 and S_2 differ by a translation along the X–axis, then it is easy to verify that

$$Z'_1 = \frac{u_1 - u_2}{X_T} Z_1^2.$$

The preceding analysis is summarized as follows:

Proposition: *If correspondence is established between two image points p_1 and p_2 of a binocular view, then the three-dimensional motion of space point P that they represent is known, given the stereoscopy parameters and the optical flow at p_1 and p_2.*

3.3 Perception of Rigid Objects in Motion

In the previous section we showed how the three-dimensional motion of points in space can be recovered from depth and optical flow. In this section we will show how the determination of rigid objects in motion can be achieved. We will use a characterization of rigid motion in the form of a theorem which relates depth, image positions, and optical flow to the motion of rigid objects in space. This theorem has been derived (Mitiche, 1984a) from a classic theorem in mechanics (Lelong-Ferrand and Arnaudies, 1974) and can be stated informally as follows:

THEOREM: Let a body, B, be defined by a discrete set of points, E. A C^1-motion of B is a rigid motion *if and only if*

$$(\forall t)(\forall M \in E)(\forall N \in E) \; aZ_M^2 + bZ_N^2 + cZ_MZ_N + dZ_MZ'_M + \qquad (3)$$

$$eZ_MZ'_N + fZ_NZ'_N + eZ_NZ'_M = 0.$$

A C^1-motion is a "smooth" motion where the functions that represent the positions of points in space as a function of time are continuously differentiable. Also, t is time; Z designates depth; Z' is the Z-component of three-dimensional velocity, i.e., the derivative of depth with respect to time; M, N have image coordinates, (x_M,y_M), (x_N,y_N), and optical flow values, (u_M,v_M), (u_N,v_N), respectively; and

$$a = -u_M x_M - v_M y_M,$$

$$b = -u_N x_N - v_N y_N,$$

$$c = u_M x_N + v_M y_N + u_N x_M + v_N y_M,$$

$$d = -x_M^2 - y_M^2 - 1,$$

$$e = x_M x_N + y_M y_N + 1, \quad \text{and}$$

$$f = -x_N^2 - y_N^2 - 1 .$$

The rigid motion theorem in the form above relates optical flow and depth explicitly to three-dimensional motion. Note that coefficients a, b, c, d, e and f are in terms of image positions and optical flow values only. The proof of this theorem (Mitiche, 1984a) is based on a result in kinematics of solids (Lelong-Ferrand and Arnaudies, 1974) which, informally, states that if a body, B, is defined by a set of points, E, in space, then a C^1-motion of B is a rigid motion if and only if

$$(\forall t)(\forall M \in E)(\forall N \in E) \ (\vec{M'}-\vec{N'}) \cdot \vec{MN} = 0 \qquad (4)$$

were $\vec{M'}$ and $\vec{N'}$ are the three-dimensional velocities of space points, M and N, respectively.

Now, if the coordinates of M and N are (X_M, Y_M, Z_M) and (X_N, Y_N, Z_N), respectively, then the projective relations allow us to write,

$$\begin{bmatrix} X_M \\ Y_M \end{bmatrix} = \begin{bmatrix} x_M Z_M \\ y_M Z_M \end{bmatrix}$$

and   (5)

$$\begin{bmatrix} X_N \\ Y_N \end{bmatrix} = \begin{bmatrix} x_N Z_N \\ y_N Z_N \end{bmatrix},$$

assuming a central projection model as in Figure 1 with the projection center at the origin and a focal length of 1. From Eq. 5, the three-dimensional velocities of M and N can be expressed as follows:

$$\begin{bmatrix} X'_M \\ Y'_M \end{bmatrix} = \begin{bmatrix} u_M Z_M + x_M Z'_M \\ v_M Z_M + y_M Z'_M \end{bmatrix}$$

and   (6)

$$\begin{bmatrix} X'_N \\ Y'_N \end{bmatrix} = \begin{bmatrix} u_N Z_N + x_N Z'_N \\ v_N Z_N + y_N Z'_N \end{bmatrix},$$

where the prime symbol indicates the time-derivative. Expanding Eq. 4 we can write

$$(X'_M - X'_N)(X_N - X_M) + (Y'_M - Y'_N)(Y_N - Y_M) +$$ (7)

$$(Z'_M - Z'_N)(Z_N - Z_M) = 0.$$

Substituting Eq. 5 and Eq. 6 in Eq. 7 and after some algebraic manipulations, we obtain the desired expression, Eq. 3, in the theorem, which completes the proof.

One can readily see that the theorem above is, in fact, a test for rigidity that will discriminate between objects in different motions. Indeed, the test is expected to fail at motion boundaries allowing the extent of rigid regions, therefore, to be delineated. More precisely, given the value of image variables, i.e., image positions and optical flow, and the value of object variables, i.e., depth and velocity in depth, corresponding to a pair of points in the image, one can quickly check if these points are indeed the projections of points on the same rigid object in space using the above theorem.

Although the theorem refers to all pairs of points of a rigid body, in practice we will assume that local rigidity implies global rigidity such that it is enough to apply the test to pairs of points in small neighborhoods.

Also, note that, because it uses both depth and optical flow, the test above should be inherently more powerful than those which use depth alone or optical flow alone. In the next section we will give examples that will illustrate this fact and show the expected capabilities of the test.

3.4 Examples

In the following examples, the camera positions differ by a translation along the X-axis. The scene of interest is illustrated in Figure 2. An object is made of a rigid collection of planar patches A, B, and C. We will refer to this object simply as, the object. A planar surface, D, plays the role of a background and will be referred to as, the background. A snapshot of this scene is taken from viewpoint O, the viewing direction being the Z-axis. Refer to Figure 3. At the instant of the snapshot, the geometric configuration is such that D is at $Z = 4m.$, A at $Z = 2m.$ and B is at $Z = 3m.$ C is not visible and does not play a role here except to form, in this example, a rigid object which is physically connected. The motion of the object is a rotation about the Z-axis (the rotational velocity vector is $(0,0,2)$), and in the first example the motion of the background is a screw composed of a translation of velocity $(0,0,1)$ and a rotation of the same

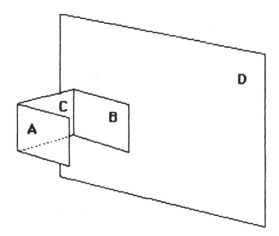

Figure 2. The observed scene.

velocity as the object.

The field of view is a portion of the image around the origin. Before attempting to segment such a scene, noise is added to depth as well as to optical flow. The magnitude and direction of optical flow have been perturbed by introducing noise independently to its u-component and its v-component. This noise is multiplicative, being uniformly distributed between $-\frac{\alpha}{2}a$ and $+\frac{\alpha}{2}a$ where a is the value to which noise is added and $\alpha = 1.5$. Noise for depth is additive and uniformly distributed in an interval centered at the value to which noise is added and of width d = 0.5m.[1] The result of applying the rigidity test,[2] as described in the previous section, is shown in Figure 4. Pluses indicate motion boundaries, i.e., points where this test has failed. The scene has been correctly segmented and we

---

[1] This amount of noise is at the limit of what the method can tolerate on this particular example without making a mistake in the segmentation. Sensitivity to noise will depend in general on the geometry of the scene and the motions involved.
[2] In these examples, the test is performed by comparing the absolute value of the left-hand side of Eq. 3 to a small threshold. We have not addressed the problem of automatic threshold selection. The test has been applied to each point and its right, lower, and lower diagonal neighbors.

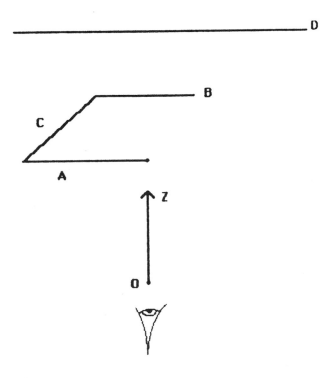

Figure 3. View of the observed scene from *O*.

can make the following important observations:

(1) If it were necessary to segment the scene from depth alone, then, because of the relationship between the depths of regions, *A*, *B*, and *D*, i.e., *B*, is at the same distance from *A* and *D*, it would not be possible to find a threshold on depth that would consistently declare *A* and *B* part of the same rigid object without also including the background *D*.

(2) As for this example, it can be shown that the contribution of translation to optical flow values is very small because of perspective around the origin. Since both the background and the object have the same rotational velocity, while their motions differ in the translational velocity, it is practically impossible to discriminate between them on the basis of optical flow alone.

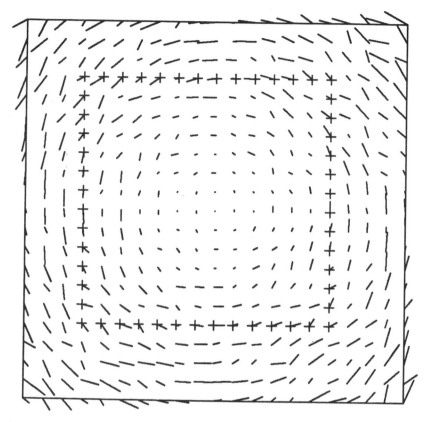

Figure 4. Optical flow and segmentation results of first example.

In the second example, everything has been kept as in the previous one except the motions involved and the amount of noise in the data. Here, both the background and the object are translating in depth with velocity vectors (0,0,2) and (0,0,4), respectively and $\alpha = 0.4$, $d = 0.2m$. Results are shown in Figure 5. As in the previous example, the scene has been correctly segmented and similar observations can be made.

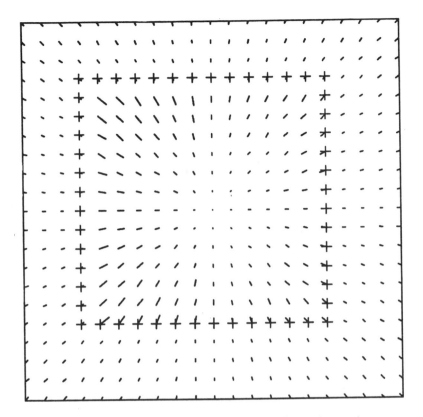

Figure 5. Optical flow and segmentation results of second example.

3.5 Summary

In this study we have addressed the problem of combining two important sources of three-dimensional information: stereopsis and kineopsis. We have described a method in which optical flow is integrated with the stereopsis mechanism to determine three-dimensional motion. The analysis allows the motion of rigid objects to be characterized and this characterization becomes the basis for a simple test that provides for the segregation of objects with different motions in space. This test, which uses both optical flow and depth, was shown by example to be more powerful than those which use optical flow alone or depth alone.

Acknowledgements

I would like to thank W.N. Martin and D.R. Proffitt for showing interest in this work. This research was supported in part by National Sciences and Engineering Research Council of Canada, under grant NSERC A6022.

References

Aggarwal, J.K., Davis, L.S., and Martin, W.N., (1982) 'Correspondence processes in dynamic scene analysis,' *Proc. IEEE,* vol. 69, no. 5, pp. 562-572.

Ballard, D.H., and Kimball, O.A., (1983) 'Rigid body motion from optical flow and depth,' *Computer Vision, Graphics, and Image Processing,* vol. 22, pp. 95-115.

Barlow, H.B., Blakemore, C., and Pettigrew, J.D., (1967) 'The neural mechanism of binocular perception,' *J. Physiology,* vol. 193, pp. 327-342.

Barnard, S.T., and Thompson, W.B., (1980) 'Disparity analysis in images,' *IEEE Trans. Pattern Analysis and Machine Intelligence,* vol. PAMI-2, pp. 333-340.

Bruss, A.R., and Horn, B.K.P., (1983) 'Passive navigation,' *Computer Vision, Graphics and Image Processing,* vol. 21, pp. 3-20.

Duda, R.O., and Hart, P.E., (1973), **Pattern Classification and Scene Analysis,** Wiley-Interscience.

Fennema C.L., and Thompson, W.B., (1979) 'Velocity determination in scenes containing several moving objects,' *Computer Graphics and Image Processing,* vol. 9, pp. 301-315.

Gibson, E.J., Gibson J.J., Smith O.W., and Flock, H., (1959) 'Motion parallax as a determinant of perceived depth,' *J. of Experimental Psychology,* vol. 58, pp. 40-51.

Horn, B.K.P., and Schunck, B.G., (1981) 'Determining optical flow,' *Artificial Intelligence,* vol. 17, pp. 185-203.

Julesz, B., (1971), **Foundations of Cyclopean Perception**, University of Chicago Press, Chicago.

Lawton, D.T., and Rieger, J.H., (1983) 'The use of difference fields in processing sensor motion,' *Proc. Image Understanding Workshop*, Washington, D.C., pp. 77-83.

Lelong-Ferrand, J., and Arnaudiès, J.M., (1974) Cours de Mathématiques - *Tome* 3, Géométrie et Cinématique, Dunod, Paris.

Longuet-Higgins, H.C., and Prazdny, K., (1980) 'The interpretation of a moving retinal image,' *Proc. Royal Society London*, vol. B 208, pp. 385-397.

Marr, D., and Poggio, T., (1979) 'A computational theory of human stereo vision,' *Proc. Royal Society London*, vol. B 204, pp. 301-328.

Mitiche, A., (1984) 'Computation of optical flow and rigid motion,' *Proc. 2nd IEEE Workshop on Computer Vision: Representation and Control*, Annapolis, MD.

Mitiche, A., (1984), 'On combining stereopsis and kineopsis for space perception,' *Proc. IEEE First Conference on Artificial Intelligence Applications*, Denver, CO, pp. 156-160.

Nagel, H.-H., (1982) 'On change detection and displacement estimation in image sequences,' *Pattern Recognition Letters*, vol. 1, pp. 55-59.

Nagel, H.-H., (1983) 'Displacement vectors derived from second-order intensity variation in image sequences,' *Computer Vision, Graphics and Image Processing*, vol. 21, pp. 85-117.

Nakayama, K., and Loomis, J.M., (1974) 'Optical velocity patterns, velocity-sensitive neurons, and space perception: A hypothesis,'

Perception, vol. 3, pp. 63-80.

Paquin, R., and Dubois, E. (1983) 'A spatio-temporal gradient method for estimating the displacement field in time-varying imagery,' *Computer Vision, Graphics and Image Processing,* pp. 205-221.

Potter, J.L., (1972) 'Scene segmentation using motion information,' *Computer Graphics and Image Processing,* vol. 6, pp. 558-581.

Prazdny, K., (1983) 'On the information in optical flows,' *Computer Vision, Graphics and Image Processing,* vol. 22, pp. 239-259.

Thompson, W.B., (1980) 'Combining motion and contrast for segmentation,' *IEEE Trans. on Pattern Analysis and Machine Intelligence,* vol. 2, pp. 543-549, 1980.

Thompson, W.B., and Barnard, S.T., (1981) 'Lower-level estimation and interpretation of visual motion,' *Computer,* vol. 14, pp. 20-28.

Waxman, A.M., (1983) 'Kinematics of image flow,' *Proc. Image Understanding Workshop,* Washington D.C., pp. 175-181.

Wohn, K., Davis, L.S. and Thrift, P., (1983) 'Motion estimation based on multiple local constraints and nonlinear smoothing,' *Pattern Recognition,* vol. 16, no. 6, pp. 563-570.

CHAPTER 4

The Empirical Study of Structure from Motion

Myron L. Braunstein

4.1 Introduction

There have been important accomplishments in the computational analysis of the recovery of information about three-dimensional relationships in the environment from dynamic two-dimensional images. As an achievement in artificial intelligence and as a significant application to robotics, these results stand by themselves. There has been increasing interest, however, in applying these analyses to the study of human visual perception. The question of applicability of a particular computational analysis to human vision is an empirical one, and one that can only be answered through rigorous empirical research. In this chapter, I will review the current status of empirical research that addresses four issues of relevance to motion understanding. The first issue is whether there are two paths to the recovery of three-dimensional structure from motion: (a) one that proceeds from the primal sketch to a viewer-centered 2 1/2 dimensional sketch to an object-centered 3-dimensional model as proposed by Marr (1982), and (b) a direct path from the primal sketch to an object-centered representation, with viewer-centered information added from separate sources. The second issue is whether the solution to the correspondence problem must precede the extraction of structure from motion. The third issue is the now familiar controversy in the perception literature about whether a rigidity constraint plays a central role in the recovery of structure from motion. The fourth issue is concerned with how the human observer distinguishes between object motion and self motion on the basis of the optical flow field. The implications for empirical research of a new theory of observers (Hoffman and Bennett, 1985), that clarifies the role of inference and of environmental constraints in perception, will be considered. Finally, a recent empirical study of shape and depth perception from motion will be discussed as an example of research directed at testing the role of specific environmental constraints in human visual perception.

4.2 Viewer-Centered vs. Object-Centered Depth

There is increasing evidence that the processes used by the human visual system to recover three-dimensional information from two-dimensional retinal images do not all recover the same type of information. In fact, one can distinguish between processes directed primarily (or initially) at recovering viewer-centered depth and those directed at recovering object-centered depth. Viewer-centered depth refers to the location of objects relative to the viewer. Distances between the viewer and the objects are recovered, although these may be either relative (correct up to a scale factor) or absolute. Near/far relationships (depth order or signed depth) are also included in this category. Object-centered depth is concerned with the shape of three-dimensional objects and with the positions of objects relative to one another, and does not include near/far relationships relative to an observer. Near/far relationships may be reversible in perceptions based only on processes that recover object-centered depth. (See Marr and Nishihara, 1978, for a discussion of the relationship between object-centered and viewer-centered coordinate systems.)

The perception of the three-dimensional environment generally involves recovery of both viewer-centered and object-centered relationships. The distinction between viewer-centered processes and object-centered processes can be thought of in terms of different paths to the recovery of three-dimensional information, as illustrated in Figure 1.

Path I. The two-dimensional information may be used to recover a viewer-centered representation of distances of feature points from the observer. An object-centered description of shapes and their spatial orientation is derived from this intermediate representation.

Path II. The two-dimensional information may be used directly in solving for a three-dimensional configuration that uniquely satisfies constraints exploited by the observer, such as rigidity. This solution may be accomplished by the visual system using the information available in orthographic projections, i.e., not using polar perspective effects. A solution based on an orthographic analysis is necessarily ambiguous with respect to relative distances from an observer, i.e., near/far relationships are not preserved. Additional visual information can be used to resolve this ambiguity.

The first paradigm (Marr, 1982) is well-suited to the analysis of stereoscopic depth processing, in which the retinal image information is initially converted to a viewer-centered depth representation. On the other hand, structure from motion analysis (Ullman, 1979) uses information in a series of two-dimensional images to recover directly an object-centered representation in which near/far relationships remain ambiguous. Indeed, disparity processing and structure from motion may be prototypical examples of processes that recover, respectively, viewer-centered depth and object-centered depth.

Consider, for example, the perception of a sphere rotating about a vertical axis. Binocular disparity can be used to recover the relative distance from the observer to each point on the sphere, i.e., viewer-centered depth. The pattern of these distances will provide the spherical shape, assuming that any scaling error is the same in all dimensions. The spherical shape is a representation of the distance relationships among the points

PATH I: VIEWER-CENTERED DEPTH RECOVERED FIRST

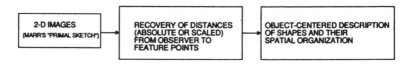

PATH II: OBJECT-CENTERED DEPTH RECOVERED FIRST

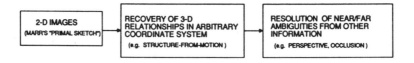

Figure 1. Two paths to the recovery of three-dimensional information
from two-dimensional images.

on the object, independent of the observer's viewpoint, and can therefore
be categorized as object-centered depth. Alternatively, the positions in the
retinal projection of points on the sphere at different instants in time can be
used, exploiting an environmental constraint such as rigidity, to solve a
system of equations that provides directly the positions of the points in
object-centered coordinates. This is the type of process that has been
labeled as the recovery of structure from motion (Ullman, 1979). For
example, it has been shown that the positions of four noncoplanar points in
three-dimensional space can be recovered uniquely in object-centered coordinates from as few as three distinct views of the two-dimensional, orthographic projections of these points (Ullman, 1979). If employed by the
human visual system, this type of process would provide object-centered
depth information directly from two-dimensional images. Viewer-centered
information could be added from other sources, such as perspective
(Braunstein, 1966, 1977b) or occlusion (Braunstein, Andersen, and Riefer,
1982; Andersen and Braunstein, 1983a). Thus, viewer-centered and
object-centered information can be recovered in both the disparity example

and the structure from motion example. However, the path to recovery of this information is reversed. Viewer-centered depth is recovered initially in the case of disparity, but object-centered depth is recovered initially in the case of structure from motion. This reversal appears to have implications for other aspects of depth perception, in particular, for size-distance invariance.

The viewer-centered vs. object-centered distinction may be related to a primitive vs. higher-order distinction, both in terms of evolution and brain physiology. Viewer-centered depth may be more primitive, automatic and inflexible, whereas object-centered depth may be more subject to cognitive influences and more characteristic of higher-level animals. This distinction, however, is not the same as that of primary vs. secondary (or pictorial) depth cues, although it appears to be related to that distinction. Looming, for example, is a form of motion perspective that has not been classified in the older literature as a primary depth cue, but is clearly among the most primitive sources of viewer-centered depth information.

The hypothesis that viewer-centered depth can be recovered directly from two-dimensional images closely parallels a conclusion concerning the direct recovery of environment-centered representations (Sedgwick, 1983). The concept of an environment-centered representation is similar to that of an object-centered representation, but places special emphasis on relationships to a continuous ground plane. Sedgwick reviews the information available for spatial layout from texture and perspective and concludes that an environment-centered representation can be recovered directly from information in two-dimensional projections, without having to pass through a viewer-centered representation.

The distinction between processes leading initially to viewer-centered depth and processes leading initially to object-centered depth is useful in interpreting two areas of current research:

a.   studies of the recovery of shape and depth information from ortho-
      graphic (parallel) projections of rotation in depth, i.e., the kinetic
      depth effect or structure from motion; and

b.   studies of the recovery of signed depth (near/far relationships) from
      velocity gradients resulting from the translation of slanted surfaces.

4.2.1 Orthographic Projections of Rotation in Depth

Human observers can recover information about the shape of objects
from orthographic projections of rotation in depth. This was demonstrated
in a classic study of the kinetic depth effect (Wallach and O'Connell,
1953) and has been verified in a number of more recent investigations
using computer-generated displays (for example, Braunstein, 1966; Braunstein, 1977a,b; Todd, 1982). The information that observers use to recover
shape from orthographic projections appears to include the sinusoidal projected velocities of rotating points (Braunstein, 1977a; Braunstein and
Andersen, 1984b), the velocity gradient parallel to the axis of rotation
(Braunstein and Andersen, 1984b), and the elliptical trajectories of points
that are rotating about oblique axes, not perpendicular to the line of sight
(Todd, 1982). In recovering shape from orthographic projections, near/far
relationships are necessarily ambiguous unless additional information is
present in the display. Occlusion of more distant parts of a display by
nearer parts is a type of information that can remove this ambiguity. This
possibility has been noted by Webb and Aggarwal (1981) and demonstrated in experiments by Braunstein, Andersen, and Riefer (1982), Andersen and Braunstein (1983a), and Proffitt, Bertenthal, and Roberts (1984).
Occlusion occurs when nearer texture elements cover more distant texture
elements, such as when texture patches on the front of a transparent sphere
cover patches on the back of the sphere. This form of element occlusion
occurs in natural settings when, for example, the closer branches or leaves
on a tree occlude the view of more distant branches and leaves. Occlusion

also occurs when an opaque surface changes in orientation relative to the observer. This is edge occlusion, sometimes called contour occlusion or self occlusion.

The implication of this research with orthographic projections is that object-centered depth information can be recovered even when there is no information present about absolute or relative distances to the observer. In fact, these projections are mathematically appropriate only when all parts of the display are infinitely distant. Although additional information, such as occlusion, can add near/far (viewer-centered) depth information, it does not appear to affect the recovery of shape (Andersen and Braunstein, 1983). Rotation in depth, displayed with orthographic projection and occlusion, thus provides information about the three-dimensional environment in the following order: (1) two-dimensional images, (2) object-centered depth, (3) resolution of near/far ambiguities. This might be expected to occur under natural conditions when viewing distances are relatively large and the scene is cluttered, allowing for occlusion.

Although an orthographic projection of an object, without additional information, does not specify the absolute or relative distances of features on the object from the observer, the observer does perceive the object at some distance and does perceive some parts of the object as closer than other parts. These perceived distances may be reversible, but distances do not reverse as easily in dynamic projections representing three-dimensional objects as in static displays such as the Necker cube. A rotating sphere, even in orthographic projection, is seldom reported to reverse. Although features on objects rotating about axes other than the line of sight will vary in their apparent distance from the observer as the object rotates, informal observation of such structure from motion displays indicate no noticeable differences in the perceived sizes of texture elements on the front and back of a rotating object. This suggests that size-distance invariance does not apply to the distances perceived on the basis of a structure from motion analysis. (These considerations were brought to my attention by Walter Gogel, personal communication, March 1985).

We have run pilot studies comparing size-distance invariance effects based on structure from motion displays to effects based on stereoscopic depth. The stimuli, shown in Figure 2, were projections of vertical cylinders. Dots (300) were randomly positioned on the surface of the cylinder.

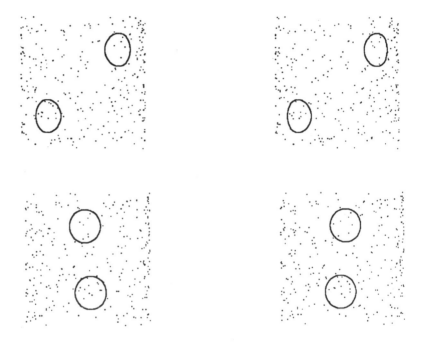

Figure 2. Two stereo pairs from a rotating cylinder sequence. Subjects judged the ratio of the sizes of the circles when they were aligned vertically in the stereoscopic view.

Two circles were drawn on the surface, separated horizontally by 180° and separated vertically. Two orthographic projections of the cylinder, rotating about a vertical axis, were displayed on a computer screen and viewed with a stereoscope. Both projections represented continuous rotations in a constant direction. A phase shift between the projections of 5° was used to induce the perception of stereoscopic depth. The subject's task was to estimate the ratio of the apparent sizes of the circles when they were aligned vertically. Viewing was either monocular or stereoscopic. In the stereoscopic condition the circle that appeared to be on the front of the cylinder was reported to appear smaller than the circle that appeared to be on the back, as expected. In the monocular condition, although a cylinder was still perceived, most subjects reported no apparent size difference for the two circles. [The cylindrical shape in the monocular condition cannot be represented in the figure as it was based on structure from motion, but we have found previously (Braunstein and Andersen, 1984b) that the shapes of cylinders and compressed cylindrical objects can be recovered accurately from orthographic projections.]

Our interpretation of these pilot results is that the perception of the cylindrical shape in the monocular motion condition is based on processes that are primarily concerned with the recovery of object-centered depth. Although relative distances (near/far relationships) are perceived, they appear to be secondary and have little if any influence on perceived size. This is contradictory to the size-distance invariance principle. If the shape is perceived as cylindrical, the perceived depth should be the same as the perceived width, regardless of the absolute perceived distance of the object. The absolute diameter of the cylinder might be perceived differently, depending on perceived absolute distance, but the ratio of the distances from the observer's viewpoint to the front and back surfaces is independent of absolute distance. This is because the absolute size of the cylinder would have to increase with absolute distance for a fixed visual angle. Size-distance invariance should therefore be independent of perceived absolute distance. [We have obtained size constancy effects in more recent pilot studies with new structure from motion displays. See Braunstein, 1986.]

The reduced effect of the size-distance invariance principle in structure from motion displays does not imply that these displays provide a cognitive interpretation rather than a perception of the three-dimensional environment. Typically at the end of our experiments subjects are asked to describe what they thought the display at which they had been looking contained. Most subjects, after viewing effective displays (Braunstein, Andersen, and Riefer, 1982; Andersen and Braunstein, 1983a), respond that they were looking at actual three-dimensional objects. The effectiveness of size-distance invariance is one possible experimental criterion for distinguishing between the two proposed paths to the recovery of three-dimensional relationships from two-dimensional images. Further research should help identify other criteria.

4.2.2 Recovery of Structure from Velocity Gradients

There have been inconsistent conclusions in the literature about the depth information available from motion parallax. Movement of the head with respect to a slanted surface produces a gradient of velocities on the retina from which it should be possible to extract both the degree and sign of surface slant. Information of this type is accurately recovered by the human observer when the velocity gradient results from active head movements (Rogers and Graham, 1979). When a velocity gradient is displayed to a stationary observer on a stationary screen, the ability of the observer to recover depth relationships is less clear.

Recovery of depth from velocity gradients has been studied using simulated dihedral angles (Farber and McConkie, 1979; Braunstein and Andersen, 1981). The angles consisted of plane surfaces bent along a horizontal line at their vertical center. In both studies, dots were distributed on the surface of the angle in a manner that produced a uniform distribution when the angle was projected onto a two-dimensional surface, so that there was no texture gradient in the projection. The angles were translated horizontally with the apex either closest to the observer's viewpoint or most distant from that viewpoint. In the two-dimensional projection of this simulated motion, the points in the vertical center of the display moved either most slowly, indicating that the apex was most distant, or most rapidly, indicating that the apex was closest. Farber and McConkie found that subjects reported perceiving a dihedral angle but responded with chance accuracy as to whether the center was nearest or most distant. Braunstein and Andersen found that accurate judgments of relative distance could be achieved, but this required longer viewing times than the 5 second durations used by Farber and McConkie. The shape of the angle, i.e., object-centered depth information, was recovered in about 3 seconds, but relative distance judgments were accurate only when the viewing time was extended to 10 seconds. This suggests that object-centered depth was recovered initially, much as in the case of rotating objects viewed with orthographic projections, but that veridical viewer-centered depth information had to be added later. This is analogous to the use of occlusion to remove the near/far ambiguity in an orthographic projection of a rotating object. The depth information in the projection of a slanted surface undergoing a horizontal translation, however, is based entirely on polar perspective. Each dot in the display moves horizontally at a constant speed, but the dot speeds vary in accordance with the dot's simulated distance from the observer's viewpoint. In an orthographic projection, which is equivalent to placing the observer's viewpoint at an infinite distance from the moving dots, all of the dots would move at the same speed and there would be no information about surface slant. A translating dihedral angle could not be perceived in an orthographic projection, as no gradient would be present.

Hoffman (personal communication, February 1984) suggested an explanation of the velocity gradient results consistent with current thinking in computational vision about the importance of orthographic projections in recovering structure from motion. According to this explanation, the visual system does not process this display as the projection of translating surfaces, at least not initially. Instead, the display is subjected to an orthographic analysis in which the velocity gradient is attributed to rotation about a vertical axis. Of course, this analysis is geometrically inappropriate as a global analysis of the projection, but applied locally, it can recover the shape of the dihedral angle. An orthographic analysis, however, could not recover the relative distance (near/far) relationships. This is illustrated in Figure 3.

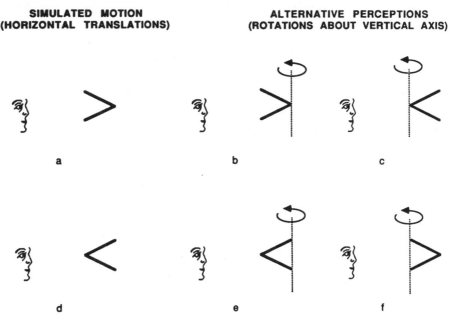

SIMULATED MOTION
(HORIZONTAL TRANSLATIONS)

ALTERNATIVE PERCEPTIONS
(ROTATIONS ABOUT VERTICAL AXIS)

a   b   c

d   e   f

Figure 3. Alternative perceptions of a translating dihedral angle. A
velocity gradient based on (a) translation with the apex pointed away
from the observer may be analyzed locally as rotation with the angle
orientation perceived (b) correctly, or (c) incorrectly. Similarly, a gradient based on (d) translation with the apex pointed towards the observer may be analyzed locally as rotation with the angle orientation (e)
correct or (f) incorrect.

The center-far condition could be interpreted as a rotation with the axis
through the apex of the angle. The orientation of the angle would be
ambiguous under that interpretation. Similarly, the center-near condition
could be interpreted as rotation about an axis through the edges of the
angle. The hypothesis that a horizontal translation is initially processed as
a rotation about a vertical axis, and this produces an ambiguous viewer-
centered interpretation, is supported by experimental results (Braunstein
and Andersen, 1981). Subjects in these experiments spontaneously
reported perceptions of rotary motion. When asked to rate amount of

rotary motion, higher ratings were obtained when the reported orientation of the angle was incorrect.

The evidence reviewed in this section is consistent with the hypothesis that there are distinct paths to the recovery of three-dimensional relationships from two-dimensional images, but it is not yet sufficient to reach a firm conclusion about the classification of perceptual processes as initially recovering viewer-centered depth or object-centered depth. Nevertheless, it should be a useful distinction to keep in mind, especially when attempting to explain discrepancies between the perceptual effects of recovering equivalent information by means of different processes.

4.3 The Correspondence Problem

In computational analyses based on comparisons of the positions of feature points in several images, a frame to frame match of feature points is necessary before three-dimensional structure can be extracted. The solution of this correspondence problem has played a major role in theoretical accounts of the extraction of structure from motion (for example, Ullman, 1979). The necessity of matching corresponding feature points across frames can be examined at different levels that fall along a local to global continuum. At the most local level, the question is whether individual feature points must be correctly matched on successive frames before three-dimensional structure can be recovered from a series of two-dimensional images. A feature point can be regarded as an individual point if it is capable of motion in the image independently of the motion of other feature points. An alternative to matching individual feature points is to match clusters of feature points, making the solution to the correspondence problem more global by some increment. If it can be shown that human observers can recover three-dimensional structure from displays in which such averaging is necessary, it would indicate that the correspondence problem need not be solved at the most local level. The necessity for correspondence matches can be questioned at a more global level. If there are no feature by feature matches possible, even with averaging across groups of points, recovery of structure would have to be based on global changes in the images from frame to frame. If this occurs in human vision, it would have to be concluded that the recovery of structure from motion is possible without any use of correspondence.

A series of experiments and demonstrations have been reported that address the correspondence issue at both the local and global level (Todd, 1985). The perceptual significance of violations of correspondence was considered for three conditions: (1) movement of configurations of points; (2) deformation of continuous contours; and (3) deformation of smoothly varying patterns of shading or texture. These experiments and demonstrations will be discussed in the next three subsections.

4.3.1 Point Configurations

The first experiment considers whether structure can be recovered when appropriate correspondence matches can be made for only a small portion of the points in a series of images. The displays consisted of 24 views of 100 dots randomly positioned on a frontal plane. The plane oscillated through an angle of 10°, 20°, 30°, 40° or 50°. Polar projection was used. The percentage of dots on two successive frames for which there was a correspondence match, appropriate to the oscillation of the plane, was 0, 12, 25, 50, or 100%. Subjects were required to judge the angle of oscillation as either 0°, 10°, 20°, 30°, 40°, or 50°. There was a high correlation between simulated slant and accuracy of judged slant at all levels of correspondence except 0%. Reduced correspondence resulted in reduced accuracy of judged slant.

One possible explanation of these results is that subjects used the static texture gradient on individual frames of the display to judge the slant, but the lack of correlation between simulated slant and judged slant at 0% correspondence, which maintains the texture gradient, showed that this was not the case. An alternative explanation is that the visual system is able to segregate signal and noise elements in the display according to the distance moved in the image. With low dot density, the mean distances among noise elements was greater than the mean distances moved by the signal elements. If a noise dot on one frame were matched to a noise dot on the next frame, using a nearest neighbor criterion, the apparent distance moved by the noise dot would, on the average, be greater than the distance moved by a signal dot. To eliminate this distinction between signal dots and noise dots, Todd generated two high-density displays. Each contained 10,000 dots on a plane moving through a 50° oscillation. One display had 100% correspondence; the other had 50% correspondence. With 100% correspondence, high density increased perceived depth in the display. With 50% correspondence, on the other hand, high density decreased perceived depth. Subjects reported perception of a rotating surface camouflaged by scintillation. This result supports the conclusion that subjects can recover three-dimensional structure from noisy displays by segregating signal and noise dots on the basis of the amount moved from frame to frame. Increasing the density of the display reduces or eliminates this distinction between signal and noise elements.

As noted earlier, if there is no correspondence solution possible at the most local level, a solution might be possible through a more global averaging process. This possibility was examined by considering constrained and unconstrained noise conditions (Todd, 1985). In the constrained noise condition, random displacements of dots from frame to frame were limited to 15 minutes of arc, or 0.2 cm. A high level of accuracy for constrained noise was found even with 0% correspondence, leading to the conclusion that an averaging process was used that included the perception of global as well as local, relative movement. Subjects reported that the constrained noise display resembled a surface with luminous ants. This is remarkably similar to the reports of subjects viewing rotating spheres and cylinders on which the points were moving at varying velocities (Braunstein and Andersen, 1984b). This may indicate a strong tendency to analyze displays into common global movements and local relative movements in the recovery of structure from motion. This would be consistent with Johansson's theory of perceptual vector analysis (Johansson, 1964).

Todd concluded from this first series of experiments and demonstrations that there are two strategies for dealing with noisy inputs: 1) isolating populations of moving elements, on the basis of the magnitude of optical motions (or perhaps some other characteristic if present); and 2) averaging motions over space/time. He points out that the averaging strategy is not necessarily incompatible with the use of correspondence matches, which could take place following a signal-averaging stage and prior to structure from motion analysis.

4.3.2 Contour Deformation

A second type of display for which the need for point by point correspondence was challenged is the deformation of occluding contours (Todd, 1985). The displays studied were opaque, filled solids rotating in depth. The actual points forming the occluding contours changed continuously, removing the possibility of correspondence matches based on individual feature points. In the first demonstration, a horizontal ellipsoid was rotated about a vertical axis. Subjects reported perceptions of a deforming two-dimensional shape, rather than rotation in depth. Similar results were obtained with a vertical ellipsoid. However, when the two shapes were shown intersecting, as in Figure 4a, subjects reported perceiving a rigid three-dimensional configuration. If the two shapes were shown adjacent but not intersecting, as in Figure 4b, the perceptions were multistable, including deforming two-dimensional shapes and rigidly rotating three-dimensional configurations. The changing pattern of intersections in Figure 4a apparently was effective in providing information about three-dimensional structure. The yoked deformation of the two shapes in Figure 4b was partially effective, but a single deforming shape was not effective. It was noted (Todd, 1985) that a correspondence-based analysis cannot

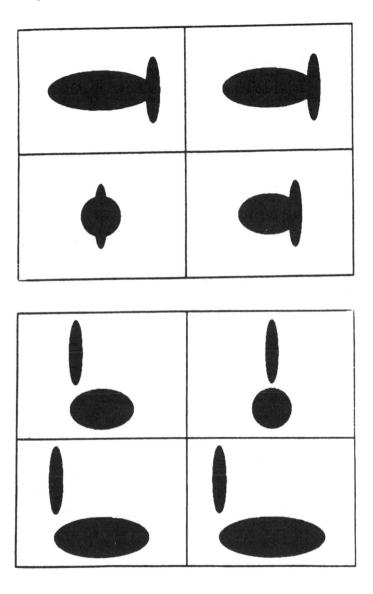

Figure 4. Static images from motion sequence depicting (a) intersecting ellipsoids, and (b) nonintersecting ellipsoids. From Todd (1985). Copyright 1985 by the American Psychological Association. Reprinted by permission of the publisher and author.

account for these results, while Koenderink and van Doorn's analysis, based on self-occluding contours (Koendrick and van Doorn, 1976), cannot account for the ability of subjects to distinguish spherical and elliptical shapes in the intersecting shape case, and cannot account for the effectiveness of yoked movement.

The effectiveness of global changes, with no local correspondence, on the recovery of structure was examined for intensity fields and texture gradients. The first case was the deformation of iso-intensity contours. The simulated object, a horizontal ellipsoid with a smoothly varying pattern of surface shading, was rotated about a vertical axis. Illumination by a distant light source was simulated. The display produced a compelling three-dimensional perception when in motion, with the shape of the solid correctly identified. Identification of the shape was not correct in static views. Todd (1985) reviews two computational analyses of this type of stimulus (Horn and Schunck, 1981; Koendrick and van Doorn, 1980, 1982). In Horn and Schunck's analysis, deformations of the intensity field are transformed into a vector field of velocities. Projective correspondence is then satisfied with respect to the transformed field. Todd points out, however, that Horn and Schunck's analysis is based on "highly restrictive assumptions." These assumptions are: (a) a flat surface; (b) uniform incident illumination across the surface; and (c) a reflectance function that varies uniformly with no spatial discontinuities. The alternative analysis is based on singularities in the pattern of image intensity (Koenderink and van Doorn, 1980, 1982). Todd points out that this analysis provides the sign of Gaussian curvature, recovering only qualitative information about the shape.

4.3.3 Texture Deformation

In the next series of demonstrations, cases are considered in which no correspondence matches are possible, even with signal averaging (Todd, 1985). One such case is a texture density gradient, which can be regenerated on each frame with no corresponding points. This was already shown to be ineffective in the first experiment described in Section 4.3.1 (Todd, 1985; and in earlier work that Todd summarizes). However, these displays considered texture density alone, without appropriate variations in the projected sizes and shapes of the texture elements. To incorporate these variations, sequences were generated in which square elements were located on the surface of a horizontal ellipsoid, covering 25% of the surface (Todd, 1985). There was no frame to frame correspondence. Subjects were not able to recover the shape of the object from static views, but when rotation about a vertical axis was simulated the reported perceptions were of a solid object with a swirling motion or scintillation on its surface. Todd points out the Koenderink and van Doorn's (1980, 1982) field structure analysis might apply to this display.

These results suggest that the human observer is not dependent on point by point solutions of the correspondence problem in the recovery of structure from motion. When appropriate correspondence matches cannot be made for all feature points, observers appear to be able to segregate signal from noise elements. When correspondence matches cannot be made at the most local level, they are made more globally. Even when there is no correspondence of specific features or groups of features, three-dimensional shape can be recovered from deformations of texture, shading, and self-occluding contours. It was concluded from these findings that the constrained boundary conditions within which existing computational models perform are not reflective of the high level of generality exhibited by human observers in perceiving three-dimensional relationships on the basis of two-dimensional images (Todd, 1985).

4.4 Rigidity

The rigidity constraint (Ullman, 1979) is related to an old idea in perceptual psychology—the principle of minimum object change. According to the principle of minimum object change, perceived changes in the size and shapes of objects should be the minimum possible, consistent with the geometry of the retinal image. This principle is easily illustrated by considering a projected square increasing or decreasing in size. The principle of minimum object change would predict that a square of constant size changing in distance will be perceived, and this is the empirical result (Johansson, 1964). Any perceived change in size would be a violation of that principle.

There are some differences in emphasis between the principle of minimum object change and the rigidity constraint. The older principle was usually applied globally to entire objects and was often applied in a top down process that used knowledge of characteristics of objects as the basis for applying the principle, although some investigators, notably Johansson and his collaborators, have used minimum object change as a bottom-up decoding principle (Jansson and Johansson, 1973). The rigidity constraint proposed by Ullman (1979) is more rigorously defined and is more local in its implications. It can be applied to a small number of points and views, i.e., three views of four noncoplanar points, in recovering three-dimensional structure from motion. Two conditions must be met in order for the rigidity constraint to be applied to the recovery of structure from motion. First, three-dimensional cues that may conflict with the structure from motion analysis should not be present in static views. The rotating trapezoid, for example, would violate this condition according to Ullman's analysis as it appears slanted in static views. The second condition is that the two-dimensional motion must not be misperceived, as might occur due to a failure in matching corresponding feature points. Such misperceptions may occur when the two-dimensional projection is a smooth continuous curve, as in the "barber pole" illusion (Hildreth, 1984).

The hypothesis that the human visual system uses a rigidity constraint in recovering three-dimensional structure from two-dimensional images has been a subject of controversy, with a number of researchers challenging this hypothesis on empirical grounds (for example, Braunstein and Andersen, 1984a; Schwartz and Sperling, 1983; Todd, 1984). A study using a rotating Necker cube was reported in which edges that represented portions of the cube closer to the observer were made brighter, as in Figure 5 (Schwartz and Sperling, 1983; Sperling, Pavel, Cohen, Landy, and Schwartz, 1983).

Figure 5. Static images from motion sequence depcting a rotating cube with conflicting luminance and perspective information for direction of rotation. From Sperling, et al. (1983). Reprinted by permission of the publisher and author.

In one situation, the cube was displayed in polar perspective with the near/far relationships represented by the perspective effects opposite to those represented by the luminance differences. Under these conditions, subjects usually reported perceiving near/far relationships in accordance with the luminance variations. The conflicting perspective effects resulted in the cube being perceived as a rubbery object with nonuniform expansions and contractions of its sides throughout the rotation. Although the geometry of the polar condition was consistent with rigid motion, any tendency to perceive rigid motion was overridden by the conflicting luminance cues. This suggests that a rigidity constraint, if used by human observers, is not a high priority principle in the recovery of structure from motion.

Research has been reported using a version of the stereokinetic effect in which a continuous curve with three self-intersections, as shown in Figure 6, was rotated in the plane (Braunstein and Andersen, 1984a). There is

Figure 6. A two-dimensional figure that appears to be a nongrid three-dimensional object when rotated in the plane. Designed by artist Fred Duncan and reproduced by permission. (See Duncan, 1975.)

a rigid interpretation of this display consistent with the retinal projection—the two-dimensional curve itself, rotating in the plane. Observers, however, reported perceiving a nonrigid, twisting and bending three-dimensional object. Ullman (1984) has commented that the results presented by Schwartz and Sperling and by Braunstein and Andersen do not contradict the rigidity principle, because the stimuli studied violated conditions required for the operation of that principle. The distorting Necker cube was considered to violate the first condition (discussed above) because three-dimensional cues were available in static views. The stereokinetic display was considered to violate the second condition on the basis that the two-dimensional motion of the curve would be misperceived. This would be expected for a figure that consisted of smooth contours lacking traceable features.

The response (Braunstein and Andersen, 1986) to these comments argues that three-dimensional structure is perceived in virtually all static views of the three-dimensional environment, except for those artificially generated in a laboratory. The first condition, then, may be too restrictive to be useful in the study of human vision. With respect to the three-looped figure, the argument is that this figure does not fall within the category of smooth contours lacking traceable features in the two-dimensional view, as the self-intersections are easily traceable. This traceability may be lost during observation of the rotating figure, but this would be a result of its three-dimensional interpretation and not a cause of misperceived two-dimensional motion.

The perception of three-dimensional structure from rigid and nonrigid motion has been studied using computer simulations of cylindrical surfaces (Todd, 1984). Each surface was defined by a pattern of 100 luminous dots, uniformly distributed on the surface. The simulated surfaces varied in curvature and were displayed using either an approximately orthographic projection or a polar projection (see Figure 7a).

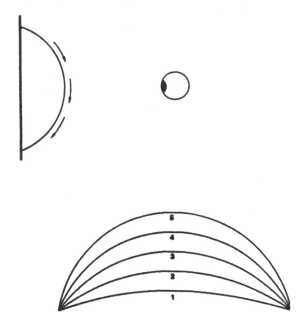

Figure 7. A section of a cylindrical surface viewed by an observer (a) and the five different curvatures (b) studied by Todd (1984). Copyright by Psychonomic Society, Inc. Reprinted by permission of the publisher and author.

Two types of surface motion were displayed, a rigid rotation and a rotation combined with stretching along the central axis of the cylinder. There were five levels of curvature displayed, and subjects were asked to judge the curvature by selecting one of five standard arcs (see Figure 7b). Judged curvature was in close correspondence to displayed curvature in all conditions. Accuracy of curvature judgments was unaffected by the rigidity of the motion. These results were related to several alternative hypotheses about the importance of rigidity in human perception (Todd, 1984). Todd concludes that the results are contradictory to an hypothesis that local rigidity is necessary in recovering structure from motion. An alternative hypothesis was discussed in which a rigidity assumption is used in specialized situations, with other more general analyses available for nonrigid motion (Todd, 1984). Todd argues that such a contingent analysis procedure would be useful to the observer if the rigidity analysis had some advantage over the general analysis. Otherwise, it would be more parsimonious to apply the general analysis to both rigid and nonrigid cases. The results with the cylinders suggest that there is no advantage gained in the recovery of structure when the display is rigid, implying that the rigidity hypothesis does not constitute a parsimonious theory of the recovery of structure from motion. [One possible advantage of the rigidity analysis, suggested by Hoffman (personal communication, October 1985) is in computational efficiency, and hence greater speed, in recovering three-dimensional structure. This would be an interesting topic for future empirical research.]

The disagreement about whether the human visual system maximizes rigidity in object perception persists. The possible limitations of the rigidity constraint in the analysis of object motion, however, may not apply to the analysis of observer motion, i.e., self motion. Although many moving objects are nonrigid, all stationary points in the environment move rigidly relative to the observer when the head or eyes move. This role of common motion in perceiving observer motion has been reviewed by Andersen (1986) and will be discussed in the following section.

4.5 Perception of Self Motion

An optical flow field projected onto the retina can result from one of three situations: (a) the observer is moving relative to a stationary environment; (b) objects in the environment are moving relative to the observer; and (c) both the observer and objects in the environment are moving. The human observer appears to be able to determine which of these situations is occurring on the basis of the flow pattern. (See Andersen (1986) for a review.) There has been very little empirical research on how observers discriminate among these situations. Gibson (1950) proposed that motion of part of the visual field is perceived as object motion, whereas motion of the entire visual field is perceived as self motion. It has been noted that this explanation does not encompass combinations of observer and object motion that result in motion of the entire visual field (Andersen, 1986).

Another proposed basis for distinguishing between object motion and observer motion is location in the visual field. Until recently, the empirical evidence indicated that a motion stimulus must extend beyond a 30° diameter region in the center of the visual field to be effective in inducing a perception of self motion. This conclusion was based on research using rotating drums, planes rotated about the line of sight, and planes translating along a vertical axis or along the line of sight. In each of these studies, a display that was found effective in the periphery was found ineffective in central vision, i.e., within 30°. These displays always represented motion of the observer relative to a two-dimensional surface. More recently, it was found that perceived self motion can be induced with projected motion limited to a visual angle as small as 7.5° (Andersen and Braunstein, 1983b, 1985). Unlike previous studies in which observer motion was relative to a flat surface, Andersen and Braunstein used displays of radially expanding flow patterns simulating forward motion of an observer through a volume filled with randomly positioned dots. The acceleration of each dot in the two-dimensional projection relative to its radial distance from the focus of expansion was determined by its simulated position in depth. Three variables were investigated: The visual angle of the display, the density of the dot pattern, and the simulated speed of observer motion. Subjects were instructed to press a button whenever they perceived that they were moving relative to the dots. The duration of reported self motion was shortest for the largest visual angle examined (21.2°) at the highest speed level. In a second experiment, a different group of subjects rated their impression of depth for the same displays. The combination of visual angle and speed that had resulted in the shortest duration of self motion in the first experiment resulted in the lowest ratings for perceived depth in the second experiment, suggesting that induced self motion in the central visual field is related to the effectiveness with which variations in depth are simulated within the inducing display. Andersen and Braunstein (1983b, 1985) concluded that the failure of previous research to find visually induced self motion in central vision was due to the lack of optic flow information for depth variations in the displays that were presented to the central field.

These results with self motion in central vision suggest that self motion can be distinguished from object motion on the basis of information in the optical flow pattern. The conditions under which self motion or object motion are perceived may be related to the availability of information for viewer-centered depth and object-centered depth. Information that provides viewer-centered depth is likely to be more effective in inducing perceived self motion, whereas information that provides object-centered depth may be more likely to result in perceived object motion (Andersen, personal communication, September 1985). This would suggest that translation along a horizontal axis, even when appropriate to a perception of self motion, may lead to a perception of object motion if an orthographic analysis is applied that attributes the projected motion to rotation about a vertical axis, as discussed earlier. The converse of this hypothesis is that the perception of a velocity gradient as being the result of self motion will lead to a viewer-centered depth analysis rather than an orthographic analysis.

As noted above, there may be a relationship between the use of a rigidity constraint and the perception of self motion. If self motion is perceived in response to a given flow field, it would be reasonable for all projected motions attributed to observer motion to be perceived as rigidly related in three-dimensional space. If this were the case, the application of a rigidity constraint would not be limited to local regions of the visual field, but could include widely separated objects in the field. If object motion is present as well, however, a rigidity constraint would not apply to the entire visual field, but an analysis based on extracting a common motion component (Andersen, 1986) would be appropriate.

Most computational analyses of optical flow do not attempt to determine whether the flow is due to object motion, observer motion, or a combination of object and observer motion. Instead, a prior assumption may be included about the source of the optic flow. One exception to this is a model (Prazdny, 1983) that decomposes the retinal velocities into translational and rotational components and then forms a description of observer motion based on the largest consistent subset of relative motion parameters in the periphery. This procedure assumes that observer motion is relative to a stationary surface. Recent findings of induced self motion in central vision with a flow field representing motion through a volume filled with randomly located dots suggest that a more general analysis will be needed (Andersen, 1986).

4.6 A Theory of Observers

Most of this chapter has been devoted to a discussion of empirical work that is relevant to theoretical developments in computational vision. There have been recent theoretical developments, on the other hand, that should be important in evaluating the usefulness of empirical work. An important formalization and extension of some fundamental concepts in the study of perception has been provided in work dealing with the inherent ambiguity in the retinal projection or in any information provided to an observer by sensory channels (Hoffman and Bennett, in press).

Statements about the ambiguity of the two-dimensional retinal projection in providing an interpretation of three-dimensional space are ubiquitous in the perception literature, but there have been some differing interpretations of the implications of this ambiguity. Gibson (1966) has argued that this ambiguity can be eliminated when dynamic information is available, i.e., a transformation of the optic array is observed. Johansson (1970) has responded that even a transforming retinal image is inherently ambiguous. If one defines stimulus ambiguity in terms of the predictability of responses made by human observers, it is possible to have unambiguous perceptions, but this is not a useful definition for studying the processes by which this predictability is achieved. If ambiguity is defined on the basis of whether properties of the three-dimensional world can be recovered from the information in the retinal image, without any further assumptions or constraints, then the retinal image is indeed ambiguous. This ambiguity is a formal property of the retinal projection and is not concerned with whether or not the responses of typical human observers are predictable. An unambiguous perception can be derived from an inherently ambiguous retinal image only through assumptions about the environment. All perceptual processing therefore must be inferential (Hoffman and Bennett, in press). It is important to note, however, that the use of inference in a perceptual process does not imply "higher-level" mental activity. Inference can be fully automatic, mechanical, implemented by a single neuron or even by a molecule. Whether a process is learned, innate, based on reasoning-like activity or based on invariants, it must use inference. The formal basis for this statement is presented in Hoffman and Bennett's paper on observer theory.

A processor (or observer) has the function of detecting a specific event (Hoffman and Bennett, in press) as in the following example: Consider a processor that detects an event called a "fly" in a hypothetical world in which the coordinates of a fly always lie somewhere along the line $X=Y=Z$ in 3-space. The processor has access only to a two-dimensional projection of this world, a projection called the observation space. However, the processor must decide whether this three-dimensional "fly" event is occurring. It is faced with the following decision table:

Observation

	X=Y	X≠Y
FLY (X=Y=Z)		
NO FLY		

Response

For X≠Y there is no problem. The processor can respond "no fly" and be correct 100% of the time. But if X=Y then there is a big problem. If the processor responds "fly" it will be incorrect with a probability of 1. This becomes clear if we consider the three-dimensional space of possible events—the configuration space. A point in the observation space for which X=Y can be the projection of any point on the plane X=Y in the configuration space. Of those points on the X=Y plane in the configuration space, the "fly" events are only those that fall on the line X=Y=Z. But a point on a plane falling on a given line is a measure zero event. The processor can only determine from the information in the observation space whether the point falls onto the X=Y plane. If it goes beyond that information, to respond that it falls on the line X=Y=Z, it will be responding that a measure zero event has occurred, and as stated above, its probability of being incorrect is 1. On the other hand. the processor could respond "no fly" when X=Y in the observation space. Since X=Y=Z is a measure zero event in the configuration space when X=Y, this response would be correct with a probability of 1. The processor could be correct all of the time, then, if it responds "no fly" regardless of whether or not X=Y in the observation space. Unfortunately, it would also be useless.

The above reasoning and formal proofs (Hoffman and Bennett, in press) show that a processor cannot be useful unless it takes advantage of regularities in the real world, i.e., environmental constraints. To continue with the example, suppose that there are no events in the real world for which X=Y unless it is also true that X=Y=Z. [This is an extreme case taken for illustrative purposes. The case of few events not meeting the constraint rather than no events not meeting the constraint may be more typical.] Then the X=Y plane in the 3-D configuration space would have high probabilities of occurrence concentrated on the X=Y=Z line. It would then be reasonable to infer that any point in the observation space for which X=Y is a projection of a point on the line X=Y=Z. Now the processor can respond "fly" when X=Y, and instead of being incorrect with probability of 1 it can be correct with high probability. The environmental constraints, which are properties of the real world and not of the processor, have made it possible for the processor to be useful.

Hoffman and Bennett present an additional example from the study of biological motion, but for the present discussion I will use one related to the perception of surface slant on the basis of velocity gradients (Gibson, Gibson, Smith, and Flock, 1959; Flock, 1964; Braunstein, 1968). Consider a processor that examines flow fields and responds that it has detected a slanted plane whenever the projected velocities are linearly related to height in the projection plane. This would be the case, for example, if a row of fence posts stretching out in depth were viewed from a moving vehicle, as in Figure 8. But the projection illustrated in that figure could be the result of an infinite number of velocity-distance combinations. For example, the points could all be at the same distance from the observer in three-dimensional space, but moving independently at velocities proportional to their heights. This might be a reasonable interpretation in a movie theater or in a psychology laboratory. In the real world however, this configuration of heights and velocities occurs only when the points are at different distances and moving rigidly with respect to the observer. The occurrence of this configuration under other conditions is possible but is a

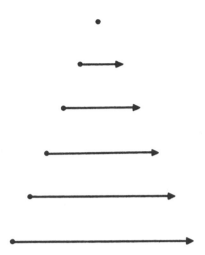

Figure 8. A velocity gradient based on observer motion past stationary objects at varying distances.

measure zero event in the configuration space. The velocity gradient processor is therefore a useful one, because it takes advantages of environmental constraints in inferring a three-dimensional layout from a two-dimensional projection.

Hoffman and Bennett's theory provides a basis for evaluating the potential usefulness of processors that might be suggested by empirical studies of motion perception. A processor should respond to an event that has measure zero in the observation space with an interpretation that would have measure zero in the configuration space, if environmental constraints were not considered. The environmental constraints, however, should ideally result in the interpretation having a high probability if the observation space event occurs. A complete theory of perception, according to this approach, must specify the measure zero events that lead to an interpretation and must specify the environmental constraints that make this interpretation useful. A theory that does not specify environmental constraints cannot be valid because it cannot handle the problem of the inherent ambiguity in sensory data.

4.7 An Empirical Test of Constraints

It is clear from the discussion above that any theory of motion under-standing that is to apply to human visual perception must incorporate the environmental constraints that are used by the human visual system. The search for relevant environmental constraints is not new to the study of visual perception. It was a major emphasis of Gibson's (1950, 1966, 1979) work. Testing the predictions of computational theories with regard to the use of environmental constraints, however, is new to perceptual psychology and presents some serious methodological problems. A number of theories are concerned with the minimum conditions required for the recovery of structure from motion under varying constraints (for example, Bennett and Hoffman, 1985; Hoffman and Bennett, 1985; Ull-man, 1979; Webb and Aggarwal, 1981). Testing theories of minimum conditions implies use of stimulus materials that present small numbers of points and views. Many perceptual psychologists, especially adherents to Gibson's principles, consider such stimulus materials to lack ecological validity and therefore to be inappropriate for research on human visual per-ception. A related difficulty exists on the response side. Recovery of structure from motion can be precisely defined as recovering the coordi-nates of all feature points in all views in an arbitrary three-dimensional coordinate system. This criterion can be implemented in a computer pro-gram, but a human subject probably could not report recovery of structure in those terms, without considerable loss of accuracy in the reporting pro-cess. A subject possibly could be trained in some task that would facilitate reporting recovery of structure, but it would then not be clear whether the constraints used were those naturally used by human observers or were artifacts of the training.

A further complication is that three-dimensional structure must be recovered from two-dimensional images. Any task given a subject must be one that can be performed on the basis of two-dimensional images, but then how do we know if the three-dimensional structure was recovered or if the subject has found some other way to relate task performance to characteristics of the images? For example, suppose that a subject was asked to determine whether or not five points shown in orthographic projection formed a rigid configuration, and suppose further that a rigid display consisted of a rotation about a vertical axis, whereas a nonrigid display contained one point rotating about an oblique axis. This task might be performed by recovering the three-dimensional structure, or simply by noting whether one of the two-dimensional paths was not parallel to the others. This potential confound is rather obvious, but there are more subtle problems of this nature that make objective psychophysical research that directly tests computational theories very difficult.

One approach that we have used is to test constraints under conditions that far exceed the proposed minimum conditions, and to use judgments of the shapes of simulated three-dimensional objects as an indirect measure of the recovery of structure (Braunstein and Andersen, 1984b). We investigated three environmental constraints: (a) Each feature point tends to move at a constant speed along its three-dimensional trajectory; (b) adjacent feature points tend to move at the same speeds (part of a rigidity constraint); and (c) overlapping fields of feature points moving in opposite directions are separated in depth. We do not believe that these constraints are strictly applied, but rather that they are applied in an approximate manner as perceptual heuristics.

We tested the first constraint by varying the shape of the three-dimensional trajectories of 160 dots simulating points on the surface of a sphere. The trajectories, which were always in a horizontal plane parallel to the line of sight, were either circular or elliptical with varying eccentricity. Each dot moved at a constant speed along its three-dimensional trajectory. An orthographic projection of these motions (125 frames) was shown to the subjects. If subjects used a constant speed constraint, or in this case a constant angular velocity constraint, in judging shape, then they should judge the three-dimensional shape in each display according to the projected velocity function. For example, a sinusoidal projected velocity function would imply a constant angular velocity circular motion in depth, and the overall shape should be judged as spherical. A projected velocity function that combined a sinusoidal and a linear component would imply motion in depth along an elliptical trajectory, with the eccentricity directly related to the relative magnitude of the linear component. The implied three-dimensional shape would then be an ellipsoid. If the projected motion were entirely linear, a flat surface would be implied.

We used the same displays to test the second constraint—that adjacent points have the same three-dimensional speeds, or angular velocities in this case. If this constraint were used by the human observer, a velocity gradient in the vertical dimension, in which the projected velocity was a function of the square root of the complement of the squared vertical position of the point, would imply curvature appropriate to a sphere. A constant projected velocity in the vertical dimension, on the other hand, would imply that the shape was flat in that dimension. These two vertical velocity gradients were examined for each of the horizontal velocity functions described in the preceding paragraph.

We tested the use of the third constraint by using displays representing either a transparent sphere or an opaque sphere. In the first case, the projection consisted of fields of dots moving in opposite directions. In the second case, all dots moved in the same direction.

Subjects were asked to rate spherical shape on an 11-point scale. They were shown a scale on which a circle was placed at the 0 position, a three-dimensional model of an ellipsoid was placed at the 5 position, and a sphere was placed at the 10 position. We found that subjects judged the shape primarily according to the horizontal waveform, indicating that the first constraint is primary in these judgments. The second constraint had a definite role, but its effect was considerably smaller. The greatest effect of the vertical velocity gradient occurred as the linear component in the horizontal waveform increased. Similarly, the effect of the third constraint was greatest when the other two types of information indicated maximum flatness. For the opaque spheres, judged spherical shape ranged from less than 2 to more than 9 on the 0 to 10 scale as a function of the first two constraints.

This type of research provides evidence about the relative importance of proposed environmental constraints, but does not test specific theories of minimum conditions. It must be supplemented by research which includes displays of the minimum numbers of points and views proposed by these theories. Such tests are currently in progress in our laboratory, using same-different judgments to measure the ability of subjects to recover three-dimensional structure.

4.8 Summary and Conclusions

In this chapter I have reviewed empirical research on four issues of relevance to motion understanding. The first issue considered was the type of analysis used in the recovery of three-dimensional structure from two-dimensional images. Some computational analyses of vision pass through a viewer-centered stage, whereas others move directly from two-dimensional images to an object-centered three-dimensional representation. Perceptual research suggests that both types of processes are used by the human visual system. The direct recovery of three-dimensional structure from two-dimensional images may be based on the information available in orthographic projections, and this may explain the lack of effectiveness of polar perspective information for recovering structure for some stimuli and the recovery of shape prior to the recovery of near/far relationships for other stimuli. Research on perceived self motion, on the other hand, indicates that viewer-centered depth is recovered from polar perspective. These considerations point to the importance of making explicit the distinction between processes that initially recover viewer-centered depth and those that directly recover object-centered depth, both in computational analyses of vision and psychophysical research.

Research on projective correspondence (Todd, 1982, 1984, 1985) indicates that point by point correspondence is not necessary for the recovery of structure from motion. The human visual system can recover three-dimensional structure by segregating signal and noise elements, by local averaging, or by using global changes in the appearance of an object over time. This suggests a complexity and flexibility in human visual processing that is not incorporated in any single computational analysis.

The role of a rigidity constraint in the perception of object motion is still in dispute, and it may be useful to distinguish between perceived object motion and perceived self motion in considering the importance of this constraint in human vision. Indeed, rigidity as a processing constraint may be related to two different environmental regularities. The preservation of approximate local rigidity during object motion is an environmental regularity based on the physical characteristics of objects and the motions that objects are likely to undergo in the environment. During observer motion, there is a preservation of global rigidity based on the preponderance of stationary objects relative to the moving observer. Most of the criticism of the generality of the rigidity constraint has been concerned with the first regularity. Future research on the use of a rigidity constraint based on the second regularity should be quite informative.

It has been demonstrated that the perception of self motion depends on the specific information in the pattern of optic flow and not simply on retinal location. Information about depth variations in the environment relative to the moving observer appears to be especially important. This research suggests that computational analyses of optic flow are important to an understanding of the ability of human observers to distinguish between object motion, self motion, and combinations of object motion and self motion. This is an area for future computational analysis and empirical research.

The work summarized in this chapter indicates considerable progress among researchers in computational vision and in perceptual psychology in communicating results and in sharing some basic concepts. The concept of separate information for object-centered depth and viewer-centered depth shows a clear influence of computational analyses on interpreting perceptual research and in guiding future empirical research. Similarly, studies of the information in optic flow that determines whether self motion or object motion is perceived have benefited from computational analyses of the recovery of depth and motion information from optic flow. The theory of observers, which formalizes the role of inference and of environmental constraints in perception (Hoffman and Bennett, in press), is an example of research based on a computational approach to vision that should serve as an important guide to future empirical research.

On the other hand, recent studies of correspondence and rigidity show a useful skepticism by perceptual researchers about the applicability of specific computational analyses to human vision. Many current analyses appear too simple and too restrictive in their assumptions to be useful in understanding the complex and highly flexible processing characteristics of the human observer. There is still a lack of direct empirical tests of specific computational theories for the recovery of structure from motion, but this is a direction in which important future progress can be expected.

Acknowledgments

This chapter includes research supported by National Eye Institute Grant EY 04553 to the author and a contract to D. Hoffman from the Office of Naval Research, Cognitive and Neural Sciences Division, Perceptual Sciences Group. I am indebted to George J. Andersen and Donald D. Hoffman for many valuable discussions of the issues presented in this paper and for helpful comments on an earlier draft. I would like to thank Lionel Shapiro and James Tittle for comments and assistance in the preparation of the manuscript.

References

Andersen, G.J., (1986) 'The perception of self-motion: Psychophysical and computational approaches,' *Psychological Bulletin,* vol. 99, pp. 52-65.

Andersen, G.J., and Braunstein, M.L., (1983a) 'Dynamic occlusion in the perception of rotation in depth,' *Perception and Psychophysics,* vol. 34, pp. 356-362.

Andersen, G.J., and Braunstein, M.L., (1983b) 'The perception of self-motion from stimulation of the central visual field,' *Investigative Ophthalmology and Visual Science,* vol. 24, (Sup. 3), pp. 278.

Andersen, G.J., and Braunstein, M.L., (1985) 'Induced self-motion in central vision,' *J. of Experimental Psychology: Human Perception and Performance,* vol. 11, pp. 122-132.

Bennett, B., and Hoffman, D., (1985) 'The computation of structure from fixed axis motion: Nonrigid structures,' *Biological Cybernetics,* vol. 51, pp. 293-300.

Braunstein, M.L., (1966) 'Sensitivity of the observer to transformations of the visual field,' *J. of Experimental Psychology,* vol. 72, pp. 683-687.

Braunstein, M.L., (1968) 'Motion and texture as sources of slant information,' *J. of Experimental Psychology,* vol. 78, pp. 247-253.

Braunstein, M.L., (1977a) 'Minimal conditions for the perception of rotary motion,' *Scandinavian J. of Psychology,* vol. 18, pp. 216-223.

Braunstein, M.L., (1977b) 'Perceived direction of rotation of simulated three-dimensional patterns,' *Perception and Psychophysics,* vol. 21, pp. 553-557.

Braunstein, M.L., (1986) 'Dynamic stereo displays for research on the recovery of three-dimensional structure,' *Behavior Research Methods, Instruments, and Computers,* vol. 18, pp. 522-530.

Braunstein, M.L., and Andersen, G.J., (1981) 'Velocity gradients and relative depth perception,' *Perception and Psychophysics,* vol. 29, pp. 145-155.

Braunstein, M.L., and Andersen, G.J., (1984a) 'A counterexample to the rigidity assumption in the perception of structure from motion,' *Perception,* vol. 13, pp. 213-217.

Braunstein, M.L., and Andersen, G.J., (1984b) 'Shape and depth perception from parallel projections of three-dimensional motion,' *J. of Experimental Psychology: Human Perception and Performance,* vol. 10, pp. 749-760.

Braunstein, M.L., and Andersen, G.J., (1986) 'Testing the rigidity assumption: A reply to Ullman,' *Perception,* vol. 15, pp. 641-644.

Braunstein, M.L., Andersen, G.J., and Riefer, D.M., (1982) 'The use of occlusion to resolve ambiguity in parallel projections,' *Perception and Psychophysics,* vol. 31, pp. 261-267.

Duncan, F., (1975) 'Kinetic art: On my psychokinematic objects,' *Leonardo,* vol. 8, pp. 97-101.

Farber, J.M., and McConkie, A.B., (1979) 'Optical motion as information for unsigned depth,' *J. of Experimental Psychology: Human Perception and Performance,* vol. 5, pp. 494-500.

Flock, H. R., (1964) 'Some conditions sufficient for accurate monocular perceptions of moving surfaces slants,' *J. of Experimental Psychology,* vol. 67, pp. 560-572.

Gibson, E.J., Gibson, J.J., Smith, O.W., and Flock, H., (1959) 'Motion parallax as a determinant of perceived depth,' *J. of Experimental Psychology,* vol. 58, pp. 40-51.

Gibson, J.J., (1950) **The Perception of the Visual World,** Houghton Mifflin, Boston.

Gibson, J.J., (1966) **The Senses Considered as Perceptual Systems,** Houghton Mifflin Company, Boston.

Gibson, J.J., (1979) **An Ecological Approach to Perception,** Houghton Mifflin, Boston.

Hildreth, E.C., (1984) **The Measurement of Visual Motion,** MIT Press, Cambridge, MA.

Hoffman, D., and Bennett, B., (1985) 'Inferring the relative 3-D positions of two moving points,' *J. of the Optical Society of America,* vol. 75, pp. 350-533.

Hoffman, D., and Bennett, B., (in press) 'Visual representations: Meaning and truth conditions,' in S. Steele and S. Schiffer (eds.), *The Second Arizona Colloquia on Cognitive Science,* University of Arizona Press.

Horn, B.K.P., and Schunck, B.G., (1981) 'Determining optical flow,' *Artificial Intelligence,* vol. 7, pp. 185-203.

Jansson, G., and Johansson, G., (1973) 'Visual perception of bending motion,' *Perception,* vol. 2, pp. 321-326.

Johansson, G., (1964) 'Perception of motion and changing form,' *Scandinavian J. of Psychology,* vol. 5, pp. 171-208.

Koenderink, J., and van Doorn, A., (1976) 'The singularities of the visual mapping,' *Biological Cybernetics,* vol. 24, pp. 51-59.

Koenderink, J., and van Doorn, A., (1980) 'Photometric invariants related to solid shape,' *Optica Acta,* vol. 7, pp. 981-996.

Koenderink, J., and van Doorn, A., (1982) 'Perception of solid shape and lay-out through photometric invariants,' in R. Trappl (ed.), **Cybernetics and Systems Research,** North-Holland, Amsterdam, pp. 943-948.

Marr, D., (1982) **Vision,** Freeman, San Francisco.

Marr, D., and Nishihara, H.K., (1978) 'Representation and recognition of the spatial organization of three-dimensional shapes,' *Proc. Royal Society of London,* vol. B 200, pp. 269-294.

Prazdny, K., (1983) 'A sketch of a (computational) theory of visual kinesthesis,' in J. Beck, B. Hope, and A. Rosenfeld (eds.), **Human and Machine Vision,** Academic Press, New York, pp. 413-423.

Proffitt, D.R., Bertenthal, B.I., and Roberts, R.J., (1984) 'The role of occlusion in reducing multistability in moving point-light displays,' *Perception and Psychophysics,* vol. 36, pp. 315-323.

Rogers, B., and Graham, M., (1979) 'Motion parallax as an independent cue for depth perception,' *Perception,* vol. 8, pp. 125-134.

Schwartz, B.J., and Sperling, G., (1983) 'Nonrigid 3D percepts from 2D representations of rigid objects,' *Investigative Ophthamology and Visual Science,* vol. 24, (Sup. 3), pp. 239.

Sedgwick, H.A., (1983) 'Environment-centered representation of spatial layout: Available visual information from texture and perspective,' in J. Beck, B. Hope, and A. Rosenfeld (eds.), **Human and Machine Vision,** Academic Press, New York, pp. 425-458.

Sperling, G., Pavel, M., Cohen, Y., Landy, M.S., and Schwartz, B.J., (1983) 'Image processing in perception and cognition,' in O.J. Braddick and A.C. Sleigh (eds.), **Physical and Biological Processing of Images**, Springer-Verlag, Berlin, pp. 359-378.

Todd, J., (1982) 'Visual information about rigid and nonrigid motion: A geometric analysis,' *J. of Experimental Psychology: Human Perception and Performance,* vol. 8, pp. 238-252.

Todd, J., (1984) 'The perception of three-dimensional structure from rigid and nonrigid motion,' *Perception and Psychophysics,* vol. 36, pp. 97-103.

Todd, J., (1985) 'The perception of structure from motion: Is projective correspondence of moving elements a necessary condition?,' *J. of Experimental Psychology: Human Perception and Performance,* vol. 11, pp. 689-710.

Ullman, S., (1979) **The Interpretation of Visual Motion**, MIT Press, Cambridge, MA.

Ullman, S., (1984) 'Rigidity and misperceived motion,' *Perception,* vol. 13, pp. 218-219.

Wallach, H., and O'Connell, D.N., (1953) 'The kinetic depth effect,' *J. of Experimental Psychology,* vol. 45, pp. 205-217.

Webb, J.A., and Aggarwal, J.K., (1981) 'Visually interpreting the motion of objects in space,' *Computer,* vol. 14, no. 8, pp. 40-46.

CHAPTER 5

Motion Estimation
Using More Than Two Images

Hormoz Shariat
Keith Price

5.1 Introduction

The idea of using three or more consecutive frames for motion analysis has been mentioned by many researchers (Ullman, 1979; Lawton, 1980; Yasumoto and Medioni, 1985). However, most approaches formulate the motion problem in such a way that they do not use the time flow information content of the image sequence optimally. The majority of techniques work with only two frames at a time, and even then, they are treated no differently than a stereo pair. Consequently, they do not make use of the fact that these frames are from a time sequence, that is, the third frame was taken T seconds after the second frame and 2T seconds after the first frame. (T being the time between any two successive frames.) As a result, most of the current approaches to motion analysis are formulated in such a way that they must treat a matched set of n features in three frames, as a set of 2n features in two frames. By doing so, they treat a three-frame sequence as two sets of two-frame sequences, and hence, under-utilize the information available to them.

On the other hand, it is shown that humans do, in fact, use this time flow information of the images on their retina to perceive motion and to segment multiple moving objects. Experiments with human perception of dot patterns have shown that the perception of a rigid body motion is extremely noise sensitive when only two frames are presented (Lappin, Doner and Kottas, 1980). Humans are also observed to experience difficulty in decomposing multiple moving bodies with different parameters when only two scenes are available to them.

Among the few who have considered more than two frames are Webb and Aggarwal (1982) who developed nonlinear equations and presented results on real data for two points in four or more frames, under the assumption of parallel projection. Their work starts with many of the same ideas presented here, but the parallel projection assumption changes the development completely.

Hoffman and Flinchbaugh (1982) developed a set of second order polynomial equations, also for the parallel projection case, which describe the motion of rigid objects.

Ballard and Kimball who have used the time flow content of three frames in the form of acceleration vectors (Ballard and Kimball, 1983). Their assumptions, however, are very restrictive and their use of this information is limited. They use accurate depth of the points, along with the optical flow, to calculate the 3-dimensional acceleration of the points. Nevertheless, they use these acceleration vectors only to calculate the direction of the axis of rotation.

Yen and Huang have developed equations to estimate rotation parameters for the special case of pure rotation (Yen and Huang, 1983). They show that if the correspondences of a single point is available over 3 frames, the rotation parameters, but not the object structure, can be determined.

Broida and Chellappa have applied the Iterated Extended Kalman Filter (IEKF) to dynamic motion equations of a long sequence of noisy images (Broida and Chellappa, 1985). Doing this increases the accuracy of the estimated motion parameters. However, for any significant accuracy increase, they require that point correspondences for 20-30 frames be available. They assume the structure of the object and its initial orientation are known (Broida and Chellappa, 1986).

Bolles and Baker also use a large sequence of images (120-130) to unify spatial and temporal analysis of image sequences (Bolles and Baker, 1985). By doing this, they avoid the problem of inter-frames feature matching at the expense of having to manipulate a large amount of data. So far, they have treated only the case where the motion of the camera is in a straight line.

In (Mitiche, Seida and Aggarwal, 1985), Aggarwal and Mitiche used iterative methods to solve the rigidity-based polynomial equations, as developed by Lawton (1980). In (Aggarwal and Mitiche, 1985), they use the correspondences of 4 lines in 3 distinct views to determine the structure and the motion parameters of the moving object. Their approach is unique since it uses the *direction* of the matched line segments. However, for convergence of the iterative solution of the nonlinear equations, good initial guesses are required.

In (Shariat, 1985) we presented a sketch of a new general method which uses the time flow information content of a more-than-two-frames sequence to solve the motion problem. In this paper we describe the general method in more detail and apply it to the case of 1 feature in 5 frames. For treatment of other cases such as "2 features in 4 frames" and "3 features in 3 frames," see (Shariat, 1986).

Due to the formulation of the solution in terms of the entire sequence, we can solve the motion estimation problem with few assumptions, namely:

1. The moving object is either rigid or is changing its shape slowly (with respect to the image sampling rate).

2. The motion of the object is either constant or is changing slowly (with respect to the image sampling rate).

3. The image sampling rate, and so the time elapsed between any two consecutive frames, is constant or is changing slowly.

When the motion of the object or its shape is changing, or the sampling rate is varying, this method simply gives the best estimate of the shape and motion parameters for the duration of the observed sequence.

Note that our definition of "constant motion" in the second assumption above is less stringent than its commonly used definition, which requires that the object not be under any external forces. To demonstrate the difference, let us consider look at the turning car image sequence used in this paper Figure 11. Even though the road is exerting a force on the car's tires, which is required for its turning motion, we consider it to have constant motion since it is presumably turning around a fixed axis with a fixed angular velocity.

The classes of motions covered by our approach include all unforced motions with single axis of rotation, as well as the above (forced) turning motion.

5.2 General Description of the Method

In Figure 1, we see the flow of control and the various modules involved in the implementation of the motion estimation process. In this section, we briefly review the function of each module.

1. *Feature Correspondence Extractor* - Given a sequence of images, this module first extracts features from each image frame, and then matches them over the entire sequence to get the point correspondences.

2. *Collision Detector* - This module estimates the translation component of motion of the moving object. It then estimates the focus of expansion, the time of collision, and the point of collision. Moreover, if an immediate collision danger exists, it generates a warning signal.

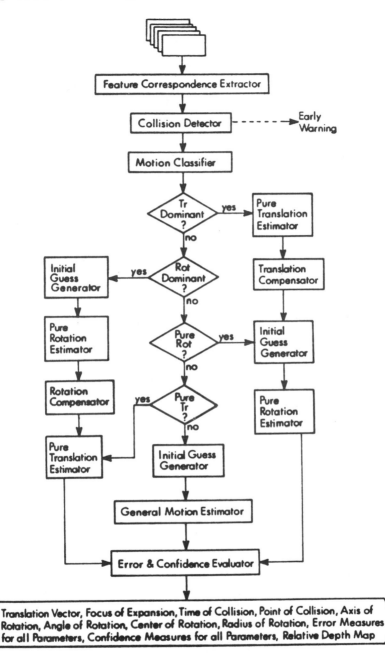

Figure 1. The block diagram of the implementation.

3. *Motion Classifier* - This is a general categorization of the motion under study. From a list of categories, the program picks one that best describes the behavior of the feature points on the image plane. The categories are: Pure Translation, Dominant Translation, Small Translation, Pure Rotation, Dominant Rotation, Small Rotation, Normal Translation and Rotation, Small Translation and Rotation, and No Motion.

4. *Pure Rotation Estimator* - This module assumes that the input data represents a purely rotational motion, and estimates the optimal rotational parameters which best fit the input data. Given some initial guesses by the "Initial Guess Generator" module, it uses the Gauss-Newton nonlinear least-square iterative method to solve a set of second order polynomial equations.

5. *Pure Translation Estimator* - This module assumes that the input data represents a purely translational motion, and estimates an optimal translation vector which best fits the input data. To do this, it solves (in a least-square sense) a set of over-determined linear equations. If several points are available, it also calculates the structure of the moving object (or the relative depth map of the environment if the observer is moving).

6. *Translation Compensator* - This module adjusts the input data to compensate for the (given) translation. When the translation component of the motion is dominant, the input data is adjusted for the translation vector that has been estimated by the "Pure Translation Estimator" module. The adjusted data represent a purely rotational motion, which is estimated by the "Pure Rotation Estimator" module.

7. *Rotation Compensator* - This module adjusts the input data to compensate for the (given) rotation. When the rotation component of motion is dominant, the input data is adjusted for the previously estimated rotation. The adjusted data represent a purely

translational motion, which is estimated by the "Pure Translation Estimator" module.

8. *Initial Guess Generator* - This module automatically generates initial guesses for subsequent iterative algorithms. Since we must solve nonlinear polynomial equations iteratively, good initial guesses are not only desired, but necessary for convergence.

9. *General Motion Estimator* - This module uses the Gauss-Newton method along with the initial guesses provided by the "Initial Guess Generator" module to iteratively solve the nonlinear general motion equations.

10. *Error And Confidence Evaluator* - An important feature of the implementation of this approach is that the program does a self-appraisal of its performance. For every estimated parameter, the following two corresponding performance measures are also given:

   a. *An Error Measure* - This is an estimate of the error in the corresponding calculated parameter. When the input data is inconsistent or extremely noisy,
   then this error measure is high, which signals the unreliability of the estimated parameter.

   b. *A Confidence Measure* - This shows how confident the program is of its estimated error measure given above. If there are more points in the input data than required, or if the points are observed for more frames than necessary, then the program has a higher confidence in its estimated error measure.

5.2.1 Establishing the Equations

The main idea used in this approach is that *if the translation of an object in space is compensated for, then every feature on the object will trace a circle in space, due to the rotation.* To give an example, assume a wheel in the x-y plane which is rolling in the positive x direction (Figure 2). Let us assume that initially (time=0) the wheel is centered at $(0,C,0)$. The rolling motion of this wheel can be thought of as the sum of a pure rotation around the axis $(0,0,1)$ located at $(0,C,0)$ and a constant, pure translation $T_x,0,0$ along the x-axis. Let us pick the point Q_0 on top of the wheel as our feature point. The location of this point at i^{th} frame (time=i-1) will be Q_i. It is quite obvious that if the total translation of this point is compensated for at each frame (i.e. if we add the vector $(-(i-1)T_x,0,0$ to it), then the compensated points, P_i's must be located on a circle.

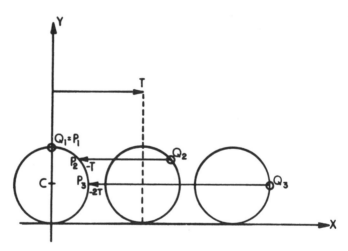

Figure 2. A Rolling wheel in the X direction when at each frame the translation is compensated for, points on the wheel will trace a circle in space.

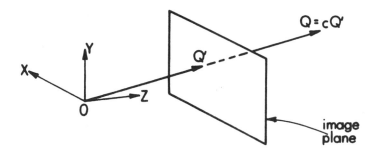

Figure 3. The image plane geometry.

Based on this fact, equations are developed, which exploit the properties of a circle in space. Doing this reduces the complex problem of
motion analysis to the problem of estimating a few coefficients from polynomial equations. Referring to Figure 3, let us denote:

Q_i The position of the feature in the ith frame in space (a 3-dimensional
      vector);

Q'_i The image of the point Q_i on the image plane (a 3-dimensional vec-
      tor);

c_i The unknown coefficient (a real and positive number) which maps
      the vector Q'_i onto the vector Q_i, so that $Q_i = c_i Q'_i$; and

P_i The position of the feature in the ith frame in space, after compensat-
      ing for the translation (a 3-dimensional vector).

The position of the feature on the third frame is arbitrarily chosen as
our reference point. So, we may write

$$Q_1 = P_1 - 2T, \tag{2a}$$

$$Q_2 = P_2 - T,$$

$$Q_3 = P_3,$$

$$Q_4 = P_4 + T, \text{ and}$$

$$Q_5 = P_5 + 2T,$$

or equivalently,

$$P_1 = Q_1 + 2T, \tag{2b}$$

$$P_2 = Q_2 + T,$$

$$P_3 = Q_3,$$

$$P_4 = Q_4 - T, \text{ and}$$

$$P_5 = Q_5 - 2T,$$

where T is the translation component of motion between any two frames.

Using the properties of a circle in space, we develop the following equations:

(I) - Since the angular velocity, ω, is constant, the distance between any two consecutive compensated points must be constant (see Figure 4). Therefore, we may write

$$\|P_5 - P_4\|^2 = \|P_4 - P_3\|^2 = \|P_3 - P_2\|^2 = \|P_2 - P_1\|^2$$

which after the substitution defined by Eq. 2b, yields

$$2T \bullet (Q_1 - 2Q_2 + Q_3) = (Q_3 - Q_1) \bullet (Q_1 - 2Q_2 + Q_3), \tag{3a}$$

$$2T \bullet (Q_2 - 2Q_3 + Q_4) = (Q_4 - Q_2) \bullet (Q_2 - 2Q_3 + Q_4), \text{ and} \tag{3b}$$

$$2T \bullet (Q_3 - 2Q_4 + Q_5) = (Q_5 - Q_3) \bullet (Q_3 - 2Q_4 + Q_5), \tag{3c}$$

where $\bullet$ denotes the vector inner product. Let us define

$$D_1 = 2(Q_1 - 2Q_2 + Q_3), \tag{4a}$$

$$D_2 = 2(Q_2 - 2Q_3 + Q_4), \tag{4b}$$

$$D_3 = 2(Q_3 - 2Q_4 + Q_5), \tag{4c}$$

$$\rho_1 = (Q_3 - Q_1) \bullet (Q_1 - 2Q_2 + Q_3), \tag{4d}$$

$$\rho_2 = (Q_4 - Q_2) \bullet (Q_2 - 2Q_3 + Q_4), \text{ and} \tag{4e}$$

$$\rho_3 = (Q_5 - Q_3) \bullet (Q_3 - 2Q_4 + Q_5). \tag{4f}$$

Eqs. 3a, 3b, and 3c may now be written,

$$T \bullet D_1 = \rho_1, \tag{5}$$

$$T \bullet D_2 = \rho_2, \text{ and} \tag{6}$$

$$T \bullet D_3 = \rho_3. \tag{7}$$

(II) - Since P_i's are on a circle and the angular velocity is constant, the angle between the vectors $(P_{i+2} - P_{i+1})$ and $(P_i - P_{i+1})$, denoted by β_i, will be the same for i=1,2,3 (Figure 4). So we may write

$$Cos(\beta_1) = Cos(\beta_2) = Cos(\beta_3).$$

In vector form this will be

$$\frac{(P_3 - P_2) \bullet (P_1 - P_2)}{(P_1 - P_2) \bullet (P_1 - P_2)} = \frac{(P_4 - P_3) \bullet (P_2 - P_3)}{(P_2 - P_3) \bullet (P_2 - P_3)} =$$

$$\frac{(P_5 - P_4) \bullet (P_3 - P_4)}{(P_3 - P_4) \bullet (P_3 - P_4)}. \tag{8}$$

Again by doing the substitutions defined by Eq. 2b and some algebraic manipulations, we get

$$T \bullet (Q_3 - Q_1 + Q_2 - Q_4) = (Q_3 - Q_2) \bullet (Q_3 - Q_1 + Q_2 - Q_4), \text{ and} \tag{9}$$

$$T \bullet (Q_4 - Q_2 + Q_3 - Q_5) = (Q_4 - Q_3) \bullet (Q_4 - Q_2 + Q_3 - Q_5). \tag{10}$$

(III) - When compensated for translation, each feature traces a circle on a plane which is perpendicular to the axis of rotation. Since the axis of rotation is unique, we may write

$$\frac{(P_2 - P_1) \times (P_3 - P_2)}{\|(P_2 - P_1) \times (P_3 - P_2)\|} = \frac{(P_3 - P_2) \times (P_4 - P_3)}{\|(P_3 - P_2) \times (P_4 - P_3)\|} =$$

$$\frac{(P_3 - P_2) \times (P_4 - P_3)}{\|(P_3 - P_2) \times (P_4 - P_3)\|} \tag{11}$$

where $\times$ denotes a vector cross product. Let us define

$$A_1 = (Q_4 - 3Q_3 + 3Q_2 - Q_1),$$ (12a)

$$A_2 = (Q_5 - 3Q_4 + 3Q_3 - Q_2),$$ (12b)

$$B_1 = (Q_4 - Q_1) \times (Q_2 - Q_3), \text{ and}$$ (12c)

$$B_2 = (Q_5 - Q_2) \times (Q_3 - Q_4).$$ (12d)

Then after substituting Eq. 2b into Eq. 11, we get two vector equations,

$$T \times A_1 = B_1, \text{ and}$$ (13)

$$T \times A_2 = B_2.$$ (14)

Note that we have

$$Q_1 = c_i Q'_i, \quad \text{for } i=1,2,3,4,5,$$ (15)

where the Q'_i are measurable from the image plane. Since the absolute depth information is lost during the imaging process, we are free to chose a scaling factor. Hence, we arbitrarily set $c_3=1$ without any loss of generality.

So, we have established 11 polynomial equations: 5 with Eqs. 5, 6, 7, 9, 10; and 6 with the vector Eqs. 13 and 14. However, these equations are not independent and may be expressed by only 8 independent equations. Note that we have 7 unknowns: c_1, c_2, c_4, c_5, and the three unknowns in $T = (T_x, T_y, T_z)$.

5.2.2 Simplifying the Equations

Before we start solving the equations, we can go another step further and eliminate T by means of algebraic manipulations. This will reduce the number of equations to 5 and the number of unknowns to 4.

First, notice that calculating the sum: {Eq. 5} + {Eq. 6} + 2*{Eq. 9} will set the left side of the sum to zero. This will eliminate T and give us a second order polynomial equation,

$$(Q_4 - Q_3 - Q_2 + Q_1) \bullet (Q_4 - 3Q_3 - 3Q_2 + Q_1) = 0. \qquad (16)$$

Likewise, from Eqs. 6, 7, and 10, we will eliminate T to get

$$(Q_5 - Q_4 - Q_3 + Q_2) \bullet (Q_5 - 3Q_4 - 3Q_3 + Q_2) = 0. \qquad (17)$$

At this point, we use the following lemma to further eliminate T from vector Eqs. 13 and 14.

LEMMA - Assume T, A, B, and C are vectors and ρ is a scalar. If we have two equations in the form $T \times A = B$, and $T \bullet D = \rho$, then the solution is

$$T = \frac{(\rho A + D \times B)}{(A \bullet D)}, \qquad provided \ (A \bullet D) = 0.$$

The proof is simple and can be done by substitutions.

We apply this lemma to Eqs. 5 and 13 to calculate T in terms of Q_i's,

$$T = \frac{(\rho_1 A_1 + D_1 \times B_1)}{(A_1 \bullet D_1)}. \qquad (18)$$

We calculate T a second time by applying the lemma to Eqs. 6 and 14, which gives us

$$T = \frac{(\rho_2 A_2 + D_2 \times B_2)}{(A_2 \bullet D_2)}. \qquad (19)$$

Finally, by equating this two calculations of T, we get

$$\frac{(\rho_1 A_1 + D_1 \times B_1)}{(A_1 \bullet D_1)} - \frac{(\rho_2 A_2 + D_2 \times B_2)}{(A_2 \bullet D_2)} = 0. \qquad (20)$$

This is a vector equation which gives us a set of 3 polynomial equations of order 5 which along with Eqs. 16 and 17 will establish a system of 5 equations with 4 unknowns which are c_1, c_2, c_4, and c_5.

5.2.3 Solving the Equations

Since in this case (and generally) there are more equations than unknowns, the Gauss-Newton nonlinear least-square method is used to solve the set of nonlinear equations. Note that the equations are in the form of polynomials. This ensures a rather good convergence of this iterative method when applied to our problem. However, since we are solving nonlinear equations iteratively, the convergence of the algorithm (that is how fast it converges to the

global minimum, if it converges at all) is highly dependent on the choice of the initial guesses provided for the algorithm.

While most methods require that *the user* supply a good initial guess, we have developed an elegant method for *automatically* generating excellent initial guesses. During application of the method described in this paper to synthetic and real images, the initial guessing module always generated valuable initial guesses. For most cases, this resulted in the convergence of the algorithm to the right answer within 20 iterations. Note that the time per iteration is 0.56 second (on Symbolics 3640) which is small especially when compared to the time required to find correspondences or calculate the optical flow.

The guessing scheme is based on the fact that the unknown coefficients c_1, c_2, c_4, and c_5 must all be real and strictly positive. These requirements are worked back into the equations to restrict the choices of initial guesses to a segment of a line. To keep this paper to a reasonable length, the details of the initial guessing scheme are not presented here and the interested reader is referred to (Shariat, 1986).

Note that since we are dealing with nonlinear polynomial equations, we should expect that multiple solutions exist. However, we have more equations than unknowns. Also, we know that the coefficients c_i's must be real and strictly positive. Theoretically, these conditions will reduce the number of acceptable solutions. Unfortunately, the current theorems in the calculus of the polynomials (on the uniqueness of the solution) are not directly applicable to our problem (mainly because of the number of equations and unknowns in our cases). However, throughout our extensive test runs (on noiseless data) we never encountered multiple answers. So our experiments support the hypothesis that the solution is unique.

5.2.4 Calculating the Motion Parameters

Once the coefficients c_i's are estimated, calculation of the translation and rotation parameters is trivial and straight forward. First, the calculated c_i's are substituted into Eqs. 18 and 19 to get two calculations for the translation vector, T, which may be averaged. Of course, this T is only proportional to the actual translation vector, T_a. The actual translation vector will be:

$$T_a = c_3 T.$$

Hence, the depth of at least one point is needed in order to get the actual translation vector.

The calculated translation vector, T, is then used in the following equations to compensate the feature points for translation and get the imaginary points P_i (which lie on a circle),

$$P_1 = c_1 Q'_1 + 2T,$$

$$P_2 = c_2 Q'_2 + T,$$

$$P_3 = c_3 Q'_3,$$

$$P_4 = c_4 Q'_4 - T, \text{ and}$$

$$P_5 = c_5 Q'_5 - 2T.$$

Looking at the plane defined by P_1, P_2, P_3, P_4, and P_5 (Figure 5), we see that $\theta = \pi - \beta$. So θ could be calculated from

$$Cos(\theta) = \frac{(P_{i+1} - P_i) \bullet (P_{i+2} - P_{i+1})}{\|(P_{i+1} - P_i)\| \ \|(P_{i+2} - P_{i+1})\|} \qquad \text{for } i=1,2,3, \text{ and}$$

$$\theta = Cos^{-1}(\theta).$$

This gives us 3 calculations for θ which may be averaged. Having θ, the angular velocity will be: $\omega = \dfrac{\theta}{\Delta t}$.

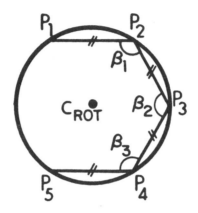

Figure 4. The plane of rotation. The translation compensated points, P_i's, must lie on a circle.

Next, we will calculate the axis of rotation, A_{Rot} ,

$$A_{Rot} = \frac{(P_{i+1} - P_i) \times (P_{i+2} - P_{i+1})}{\|(P_{i+1} - P_i) \times (P_{i+2} - P_{i+1})\|} \qquad \text{for } i=1,2,3.$$

This will give us 3 calculations for A_{Rot} which may be averaged.

Having calculated A_{Rot}, the center of rotation, C_{Rot}, is calculated. This is the 3-dimensional location of A_{Rot}. Referring to Figure 6, we use the fact that C_{Rot} lies on the bisector of the line segment connecting P_i to P_{i+1}. That is why we can write

$$C_{Rot} \bullet A_{Rot} = P_i \ A_{Rot} \qquad \text{for } i=1,2,3,4,5, \text{ and}$$

$$(P_{i+1} - P_i) \bullet C_{Rot} = (P_{i+1} - P_i) \bullet (P_{i+1} + P_i) \,/\, 2 \qquad \text{for } i=1,2,3,4.$$

This gives us a set of 9 linear equations for the 3 unknowns in

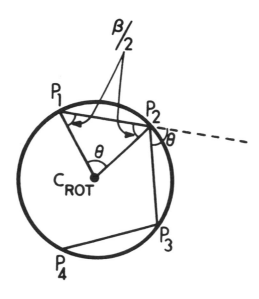

Figure 5. Calculation of the amount of rotation, θ.

$C_{Rot} = (C_x, C_y, C_z)$. They are solved for C_{Rot} by a linear least-squares method.

Finally, the radius of rotation, R_{Rot}, is calculated by

$$R_{Rot} = \|P_i - C_{Rot}\| \qquad \text{for } i=1,2,3,4,5.$$

This yields 5 calculations for R_{Rot} which may be averaged.

5.2.5 Advantages of this Approach

The advantages of this approach are listed below:

1. *Complexity Reduction* - The motion problem is reduced to one of estimating a few coefficients from polynomial equations (mostly second order). From these coefficients, the estimation of the translation vector, rotation parameters, and the structure of the moving object (the depth map of the environment, if the observer is

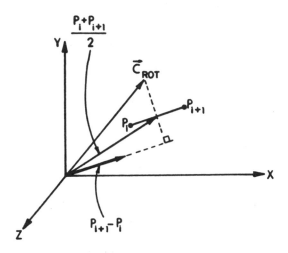

Figure 6. C_{Rot} lies on the perpendicular bisector of the line segment connecting P_i to P_{i+1}.

moving), is straightforward.

2. *Intuitive (Natural) Description of Motion* - Due to the formulation of the problem,

   a. the calculated translation vector will automatically represent the translation of the center of rotation of the object.

   b. the calculated center of rotation will automatically be the natural center of rotation.

To understand our approach, look again at Figure 2. The wheel in three frames rolling towards the positive x-axis stays in x-y plane. Intuitively, its natural center of rotation is the center of the (moving) wheel. This method develops equations with respect to the (then unknown) center of rotation. As a result, once these equations are solved, the calculated translation vector and center of rotation both assume their natural values. Describing the motion of an object in terms of natural motion parameters is important not only because that description matches our intuition, but also because natural motion parameters *are essential* for predicting the position of features in the subsequent frames.

3. *Automatic Generation of Initial Guesses* - Good initial guesses are generated automatically, which ensure a fast convergence to the right answer for most cases.

4. *Large Rotation Capability* - While most methods are formulated for small rotations, and fail if the amount of rotation exceeds a few (3-6) degrees (Fang and Huang, 1984; Adiv, 1985), this approach works increasingly better for large rotations. Its accuracy is only a function of the accuracy of the input data.

5. *Slower Sampling Rate* - Since the method works better for larger rotations and translations, it allows us to increase the interval between capturing frames. This reduces the load on the early processing modules and increases the speed and feasibility of the overall system.

6. *Structure from Rotation* - While most approaches are unable to calculate structure if no translation is present (Adiv, 1985), this method can calculate the structure under all modes of motion.

7. *Robustness* - Since a least-square solution is always sought, the method is robust under a reasonable amount of noise in the input data.

8. *Self-Appraisal* - The program generates an error measure and a confidence measure associated with every estimated parameter, to evaluate its performance and the accuracy of the results.

9. *Options in the Input Data* - The program is flexible in the number of point correspondences and frames needed. The user has a choice of 3 points in 3 frames, 2 points in 4 frames, or 1 point in 5 frames.

10. *Early Warning of Collision Dangers* - The trajectory of approaching objects are initially estimated to warn the system of any collision danger.

11. *Classification of Motion* - The program diagnoses the general mode of motion very early in the program.

5.2.6 Limitations of Our Approach

1. *Slow Rate of Convergence for Small Rotations* - Currently the convergence of the program is relatively slow for rotations smaller than 10° per frame. Although most cases converge within 20 iterations, the case of small rotation usually requires 30-70 iterations for convergence (up to a few hundred iterations for some cases). Approximate time per iteration is 0.56 seconds on the Symbolics 3645 Lisp Machine (with the garbage collection on). By generating better initial guesses and/or using a more sophisticated iterative technique, it might be possible to improve the convergence rate for small motion cases.

2. *Interpretation of Small Rotations as Translations* - When the rotation is smaller than 3° per frame, our method interprets it as a small translation and gives a translation vector which best approximates the rotational motion. This, however, is not a major problem because
   a small rotation is mathematically equivalent to a small translation. Furthermore, when the noise is more than 3% of the average disparity, the recovery of such a small rotation is difficult (if not impossible) anyway.

5.3 Results

We have applied the method described in the previous chapters to a number of computer generated test sequences and a few real image sequences. The test data covered the complete range of possible motions from pure translation to pure rotation, and included both small and large motions.

On the average, the program converged to the right answer for most noiseless data within 10 iterations. For noisy data, the answers were reasonable and their accuracy was directly a function of the amount of "Relative Noise" in the input data. "Relative Noise" is defined as

$$Relative\ Noise = \frac{The\ amount\ of\ noise\ in\ the\ disparity\ vector}{The\ length\ of\ the\ disparity\ vector}.$$

The only case where the convergence was slow was that of small rotation which usually required 30-70 (and sometimes up to a few hundred) iterations for convergence. The average time per iteration for the case of 1 feature in 5 frames was 0.556 seconds and for the case of 1 feature in 6 frames was 0.843 seconds on the Symbolics 3640 (with the garbage collector on). Note that these times are short, especially when compared to the time required for feature extraction and matching (or calculating optical flow).

For synthetic test data, all measurements and calculations are given in terms of multiples (or fractions) of the focal length, which is assumed to be 1. For the real data, the focal length is 50 millimeters and all measurements and calculations are in millimeters. Note that for the real cases, the dimensions of the images (given in x and y directions) are 36 × 24 mm. (for the "Turning Car" sequence), and 36 × 36 mm. (for the "Road" sequence). Also, notice that in *all* cases, the calculated motions are scaled down by the camera's focal length. (This was arbitrarily chosen since the absolute depth information--the scaling--is lost during the imaging process.)

5.3.1 Synthetic Test Data

In this section, we first choose a set of test cases, and then study the results of the treatment of these cases by the program. It is intended that the results given in this section serve only as *examples* of the behavior of the program under various types of motion.

The set of synthetic test data is chosen so that every option of the diagram in Figure 1 is exercised. Three test cases are reported on here that test the *General Motion* category.

1. Case #1 - An "easy" case, where both the translation and rotation components are large.

2. Case #2 - A "typical" case.

3. Case #3 - A "worst" case, where both the translation and rotation components are very small. This case was chosen so that various sources of errors could be identified and their effect could be studied.

Table 1. lists the parameters for each of these cases. The direction of the axis of rotation, which has unit length, is given by the angles γ and λ defined in Figure 7. The relative noise is caused by the (simulated) digitization process on the image plane, and represents the *average* noise of the disparity values. D_{avg} is the average magnitude of disparity vectors given in pixel units.

For each of these cases, the corresponding motion parameters were used to generate an imaginary moving point in space. Then, the image of this moving point on the image plane was sampled to get our test data. This formed our "ideal" (noise free) test data. Finally, a simulated digitization noise was added to each of these test data points. The added noise corresponded to the digitization of points in a camera with a unit focal length, 60° field of view, and 512×512 pixels on its image plane.

Table 2 contains the results of running the above 3 cases without any noise in the input data (i.e., without digitization noise). The second

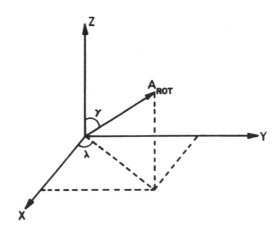

Figure 7. Definition of angles: λ and γ.

	Test Case #1	Test Case #2	Test Case #3
Motion	Large Tr & Rot	Typical Tr & Rot	Small Tr & Rot
T	(1, 2, 3)	(2, 2, 2)	(0.01, 0.02, 0.03)
C_{Rot}	(0, -1, 2)	(1, 1, 3)	(5, 5, 10)
γ	30°	45°	60°
λ	30°	45°	60°
θ	45°	20°	5°
R_{Rot}	1	3	1
Rel. Noise	3.1%	0.7%	14.8%
D_{avg} (pixels)	65.7	181.96	4.94

Table 1. General Motion Test Cases

column shows the output of the "Motion Classifier" module. As the table indicates, most cases converge to the correct answer quickly, except those with small rotations.

Case #	Motion Classification	Iterations	Run Time (Sec.)	Errors (%)
1	Normal Tr & Rot	7	3.84	$< 10^{-5}$
2	Normal Tr & Rot	14	7.79	$< 10^{-5}$
3	Small Tr & Rot	24	13.35	$< 10^{-3}$

Table 2. Results of Noise-Free Test Cases

Note that the classifications are generally, but not exactly, correct. This is because we have tuned the parameters of the "Motion Classifier" so that some amount of noise in the input data is expected (i.e. it is tuned for real scene images). Hence, for noise-free data some (minor) misclassification resulted.

General Motion Test Cases

This section reports on the general motion results for the simulated digital data (with the errors from the "digitization" process). Table 3 has the results for when 5 and 6 frames are observed. It includes both the calculation errors and number of iterations needed for convergence. The table also gives the error and performance measures generated by the program for each test case.

Figures 8. through 10. compare the calculated 3-dimensional motions with the input data for test cases #1, #2, and #3 (for only the case of 5 frames). The tags (arrows) denote the location of the input data (hollow circles) at the time of each frame. The solid curves represent the calculated 3-dimensional motion after it is interpolated and projected back onto the image plane.

As Table 3 shows, even though cases #1 and #2 have small errors, case #3 has large errors. This is due to the fact that the motion (and especially the rotation) is very small, which causes the following problems:

	Frames	Test Case #1	Test Case #2	Test Case #3
T (Length)	5	3.25%	3.71%	225.2%
T (Direction)	5	0.91°	9.87°	121.0°
A_{Rot} (Direction)	5	11.35°	52.64°	42.27°
θ	5	0.96°	5.47°	81.47°
C_{Rot}	5	0.62%	17.94%	9.26%
R_{Rot}	5	4.5%	43.67%	97.89%
Iterations	5	4	3	430
T (Length)	6	2.49%	1.09%	434.2%
T (Direction)	6	0.63°	10.8°	20.91°
A_{Rot} (Direction)	6	8.35°	53.1°	118.1°
θ	6	0.56°	8.28°	114.1°
C_{Rot}	6	0.44%	21.76%	9.14%
R_{Rot}	6	1.46%	54.66%	98.51%
Iterations	6	3	2	200
Error Measure		10^{-10} -0.2	10^{-8} -2.25	60-124
Confidence Measure		71-94	71-93	70-95

Table 3. Error Results and Performance Measures of General Motion Cases

1. T is so small that its information is lost in noise.

2. Since axis of rotation is calculated by computing the cross product of two vectors, it is sensitive to the noise in those two vectors. Thus, if the relative noise in each of those two vectors is high, then the noise in the axis of rotation is even higher.

3. Since the disparity vectors on the image plane are small (their average is 4.9 pixels), the relative image digitization error becomes high (for this case it was 14.8% of the average disparity).

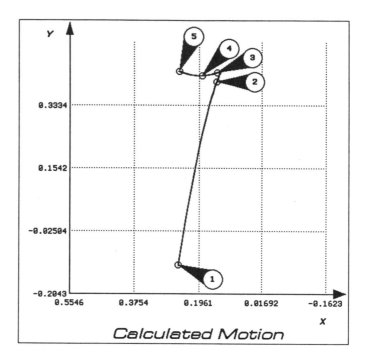

Figure 8. Calculated 3-dimensional motion for Case #1 (5 frames observed).

4. Because the disparities are small, the matrices used for the Gauss-Newton solution scheme become nearly singular, which cause large errors during their inversion.

5. The round-off errors of the calculations (i.e. the precision of the computer) start having deteriorating effects.

From this table 3, we may also see that observing an extra frame (the 6th frame) does improve the results for case #1, but not for cases #2 and #3. This shows that observing more points or more frames does not necessarily improve the results. If the additional data introduces more noise than useful information, then we expect that the results will become more

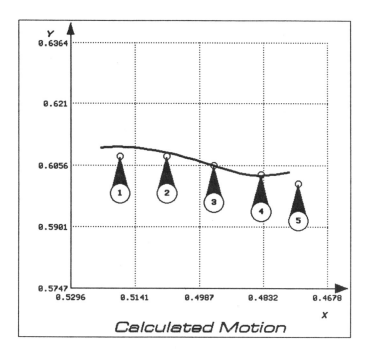

Figure 9. Calculated 3-dimensional motion for Case #2 (5 frames observed).

erroneous.

Note that according to the Table 3, the program has accurately assessed its performance. For example, it gave small error measures for test cases #1 and #2 to show the accuracy of the results. However, it gave high error measures along with high confidence numbers for test case #3. These show that the program is very "confident" that the results are erroneous. Finally, note that because of the proximity of the generated initial guesses to the solution, very few iterations were needed for convergence (in cases #1 and #2).

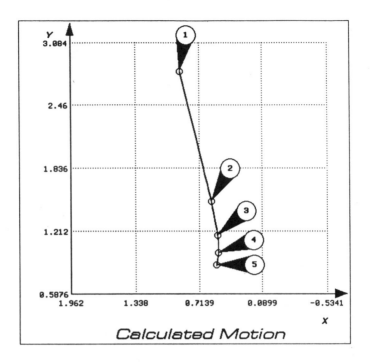

Figure 10. Calculated 3-dimensional motion for Case #3 (5 frames observed).

5.3.2 Real Test Data

In this section a few sequences of real test images are treated. In all of these cases, because the exact motion of the object is unknown, we can only *intuitively* determine if the answers are correct. However, since in each of these cases several points on the object independently give similar results, we may conclude that the answers are at least consistent.

Turning Car Sequence

Figure 11 shows 6 frames of a turning car sequence. The average disparity in this sequence of images is 25.72 pixels long. Hand-picked data as well as region-based matched data from this sequence were analyzed. The results are given below.

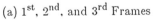

(a) 1st, 2nd, and 3rd Frames   (b) 4th, 5th, and 6th Frames

Figure 11. First 6 frames of turning car sequence.

Hand-Picked Data

Figure 12 shows the results of considering 5 frames at one time. Four features (corners) were hand-picked and used as input to the program. In this figure, the hollow circles indicate the input points and the solid curves are the calculated 3-dimensional motions projected back onto the image plane for comparison.

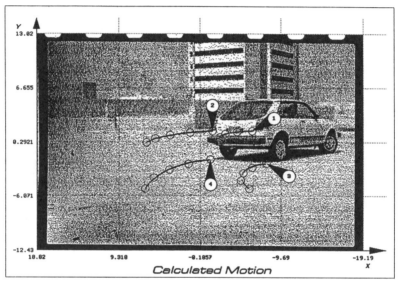

Figure 12. Comparison of input data (hollow circles) to calculated motion (solid curve)--5 frames observed.

Table 4 has the results for features #1 and #3 for both the 5 and 6 frames cases. Table 5 has similar results for features #2 and #4. The error and confidence measure for each case and each feature is given in Table 6.

As we may see in Figure 12, features #1 and #3 strongly indicate a rotation: they lie on elliptic curves. This is why their motion is detected as pure rotation (T=0) with an axis of rotation pointing down toward the -Y axis (Table 4).

	Frames	Feature #1	Feature #3
T	5	(0, 0, 0)	(0, 0, 0)
A_{Rot}	5	(-0.039, -0.98, 0.22)	(-0.25, -0.82, 0.52)
θ	5	29.1°	25.683°
C_{Rot}	5	(-6.21, 0.88, 46.78)	(-9.51, -4.58, 49.83)
R_{Rot}	5	3.85	4.47
Iterations	5	19	19
T	6	(0, 0, 0)	(0, 0, 0)
A_{Rot}	6	(-0.16, -0.97, 0.18)	(-0.026, -0.98, 0.22)
θ	6	19.27°	18.95°
C_{Rot}	6	(-6.46, 0.47, 48.32)	(-13.52, -4.12, 49.44)
R_{Rot}	6	4.08	8.42
Iterations	6	10	9

Table 4. Results of Turning Car Sequence (Features #1 and #3)

Features #2 and #4, on the other hand, have weaker indications of rotation: they lie on curves which resemble lines more than ellipses. This is logical because these two features trace a larger circle in space than the features #1 and #3 do. For this reason, a small translation component is calculated for feature #2. For feature #4, no translation is detected but the calculated rotation has a large radius of rotation (23.62 mm, measured on the image plane)--which is another way of representing near translations. For both features, the axis of rotation is accurately calculated as pointing toward the -Y axis.

The error and confidence measures given in Table 6 help the user to pick more reliable motion calculations. For example, if only 5 frames are observed, the user may want to use the motion calculations provided for

	Frames	Feature #2	Feature #4
T	5	(-1.35, 0.21, 0.32)	(0, 0, 0)
A_{Rot}	5	(-0.021, -0.99, 0.16)	(-0.16, -0.67, 0.72)
θ	5	32.91°	12.69°
C_{Rot}	5	(0.90, 0.91, 48.58)	(-4.46, -10.98, 57.99)
R_{Rot}	5	2.62	12.69
Iterations	5	1	3
T	6	(0, 0, 0)	(0, 0, 0)
A_{Rot}	6	(-0.11, -0.94, 0.31)	(-0.021, -0.995, 0.095)
θ	6	8.57°	13.30°
C_{Rot}	6	(-2.97, -1.73, 56.24)	(-18.22, -3.68, 45.97)
R_{Rot}	6	9.54	23.62
Iterations	6	9	3

Table 5. Results of Turning Car Sequence (Features #2 and #4)

	EM/CM (5 Frames)	EM/CM (6 Frames)
Feature #1	6.5/82.7	33.9/57.4
Feature #2	0.69/86.7	27.1/64.4
Feature #3	36.4/57.8	22.0/81.7
Feature #4	40.3/50.7	5.9/91.7

Table 6. Error and Confidence Measures for Features from Turning Car Sequence

features #1 or #2. However, if 6 frames are observed, the user should pick the results for features #3 or #4.

It must be pointed out that because of the good initial guesses, all cases have converged to the right answer within 20 iterations.

Region-Based Matched Data

To determine point correspondences, we used the methods described in (Ohlander, Price and Reddy, 1978; Faugeras and Price, 1981) to segment the image into regions and then match the regions on the basis of their shape, size and position. Figure 13. shows the input data along with the calculated 3-dimensional motion projected back onto the image plane for comparison. Table 7 has the calculated results.

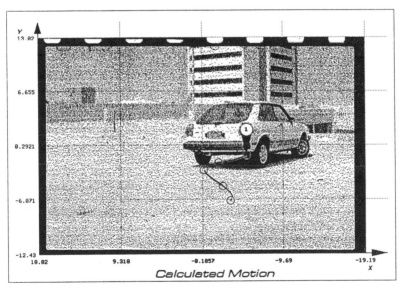

Figure 13. Comparison of input data (hollow circles) to calculated motion (solid curve)-- matched data.

In Table 7, notice that even though the axis of rotation seems to be correct (pointing toward the -Y axis), the calculations of the translation

	Feature #1
T	(-0.59, 1.36, 0.25)
A_{Rot}	(0.18, -0.97, -0.14)
θ	86.19°
C_{Rot}	(-2.56, -3.11, 50.85)
R_{Rot}	2.30
EM/CM	$< 10^{-11}/87\text{-}98$
Iterations	91

Table 7. Results of Turning Car Sequence (Matched Feature #1; 5 Frames Observed)

vector and θ seem to be wrong. This, however, is completely caused by the erroneous input data for the following two reasons:

1. As Figure 13 shows, the input data does not fit into an ellipse, and therefore does not support an estimate of pure rotation.

2. The input data along with the calculated 3-dimensional motion parameters perfectly fit into the motion equations (the residual is < 10^{-11}). So, as far as the program can decide, it has accurately found the 3-dimensional motion parameters. This is why the program has given a small error measure along with a large confidence measure (Table 7).

Road Sequence

Figures 14 and 15 show 5 frames of a road scene used as the input image sequence. In this sequence, both the observer and the other car on the road are moving. The average length of the disparities is 28.86 pixels. Figure 16 displays the input (hand-picked) data (hollow circles) versus the calculated 3-dimensional motions (solid curves) for 3 frames. Four features were hand-picked: two from the side of the road, which are used to determine the observer's motion; and two from the approaching car and its shadow, which are used to calculate the 3-dimensional motion of the other car.

Table 8 contains the results for features #1 and #2 (on the side of the road) for 3 and 5 frames cases. Table 9 has similar results for features #3 and #4 (on the approaching car, and on its shadow). Table 10 lists the error and confidence measures for both of the above cases.

The results for features #1 and #2 correspond to the motion of the observer, and results for features #3 and #4 correspond to the *relative* motion between the observer and the approaching car. The motions of features #1 and #2 are somewhat similar, even though the motion parameters for feature #2 seem to be erroneous. The resemblance and consistency of the calculated 3-dimensional motion for features #3 and #4, however, are remarkable.

By comparing the given parameters, we can make some interesting conclusions (which also give us some assurance about the accuracy of the results).

1. By comparing the translation vectors, we see that features #3 and #4 are moving about twice as fast as features #1 and #2 (especially feature #1).

2. By comparing the POC and TOC values for these features, we conclude that features #1 and #2 will pass on the far right of the observer in 13.4 and 7.8 inter-frame time units respectively. Features #3 and #4, on the other hand, will pass on near left of the

Figure 14. First 4 frames of road sequence.

Figure 15. Last frame of road sequence.

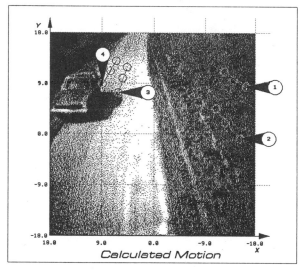

Figure 16. Comparison of input data (hollow circles) to calculated motion (solid curve)--3 frames observed.

observer in 6.3 and 6.2 inter-frame time units respectively.

3. Features #3 and #4 have similar FOE vectors (especially for the case of 5 frames). This alone (ignoring everything else) is sufficient to deduce that these two points belong to the same object. Note that even though the X-components of FOE's are

	Frames	Feature #1	Feature #2
T	3	(-1.12, -2.16, -3.62)	(-0.85, -3.28, -6.23)
FOE	3	(15.51, 29.82)	(6.81, 26.33)
TOC	3	13.81	8.03
POC	3	(-32.01, -21.34)	(-22.19, -27.08)
Run Time (sec.)	3	0.064	0.068
T	5	(-0.96, -1.86, -3.26)	(-0.66, -2.59, -5.12)
FOE	5	(14.74, 28.65)	(6.45, 25.26)
TOC	5	13.36	7.76
POC	5	(-27.19, -17.47)	(-17.40, -20.61)
Run Time (sec.)	5	0.27	0.20

Table 8. Results of Road Sequence (Features #1 and #2)

different for these four features (implying different motions), their Y-components are similar, which imply that these features are on about the same Y-plane.

As Table 10 shows, the program has signaled some errors for all of the cases. But as we stated before, the measures given for the case when 5 frames are considered are more reliable. It is also interesting to note that for features #1 and #2, using 5 frames (instead of 3 frames) has not decreased the error measure, rather, it has increased the confidence in these error measures.

Finally, it must be noted that the run time for these cases is always a fraction of a second, which is important, especially if the approaching car was on a collision path with the observer.

	Frames	Feature #3	Feature #4
T	3	(0.15, -3.82, -6.58)	(0.25, -3.87, -7.09)
FOE	3	(-1.12, 29.02)	(-1.75, 27.34)
TOC	3	7.60	7.05
POC	3	(7.20, -21.59)	(10.53, -18.34)
Run Time (sec.)	3	0.066	0.083
T	5	(-0.19, -3.12, -6.00)	(-0.14, -3.15, -6.12)
FOE	5	(1.62, 26.01)	(1.16, 25.685)
TOC	5	6.33	6.17
POC	5	(2.95, -14.08)	(5.29, -12.7)
Run Time (sec.)	5	0.20	0.26

Table 9. Results of Road Sequence (Features #3 and #4)

	EM/CM (3 Frames)	EM/CM (5 Frames)
Feature #1	9.24/62.3	9.24/90.03
Feature #2	5.41/66.1	5.34/92.0
Feature #3	1.72/69.7	5.0/92.1
Feature #4	0.85/70.5	5.37/92.0

Table 10. Error and Confidence Measures of Features from Road Sequence

5.4 Comparison with Other Methods

In this section we compare our method with:

1. The method described in (Tsai and Huang, 1981; Tsai and Huang, 1986). We call this method the "Linear" method.

2. The method described by Adiv in (Adiv, 1985). We label this method as "Adiv."

3. The method described in (Yasumoto and Medioni, 1985). We call this method the "Regularization" method.

Table 11 compares the typical errors resulted in each of these methods when the same percentage of error is present in the input data.

	Linear	Adiv	Regularization	Our Method
T (Direction)	69.55°	57.64°	11.04°	0.78°
$\|T\|$ (Velocity)	70.8%	2.14%	15.47%	2.53%
A_{Rot} (Orientation)	54.03°	78.85°	9.09°	8.99°
θ	24.13°	24.00°	0.61°	1.09°

Table 11. Comparison of Calculation Errors in Different Methods (the input error was in the range of 0.35 to 0.9 percent)

From the above table and by observing the performance of these methods under various conditions, the following statements may be made:

1. The "Linear," "Adiv," and "Regularization" methods assume small rotations. Their performance starts to deteriorate when the rotation is more than 2°. They usually fail for rotations greater than 5° or 6°.

2. Even for rotations smaller than 2°, the performance of methods such as "Linear," "Adiv," and "Regularization" greatly deteriorates when the amount of the input noise becomes greater

than 2-3%.

3. Our method makes no assumptions about the amount of rotation. However, since the calculation of the rotation parameters is based on the cross product of the vectors, the error in its calculation of the axis of rotation increases significantly as the amount of rotation falls below 3°. Nevertheless, most often such a small rotation is detected as "small translation" by our program and a translation vector is given which best describes the motion of the point. This is a logical solution to the problem of "small rotation" for our method because mathematically, such small rotation can be approximated (rather accurately) by a small translation.

4. As Table 11 indicates, our method works very well in terms of the calculation accuracy of the translation parameters (direction of T and the velocity). Furthermore, its calculation accuracy of the rotation parameters (axis of rotation and the amount of rotation) is better than "Linear" and "Adiv," and is comparable with "Regularization."

5.4.1 Error Analysis

Three of the synthetic test cases were used in an extensive stochastic error analysis. Each point was shifted by a random amount and the motion estimation program was run on the altered data. The new results (rotation and translation parameters) were compared with the original parameters and the differences in these values were stored. We will not report on all the specific details here, but will only give the general conclusions:

1. All modules perform reasonably well with noisy input data.

2. The calculated rotation parameters, especially the axis of rotation, is more sensitive to noise than the translation parameters. This is caused by the fact that the rotation parameters involve the use of vector cross products which are more affected by small changes in

the input vectors.

3. Regardless of which module is used for the calculation of T ("Pure Translation Estimator" or "General Motion Estimator"), the calculation of T is reasonably accurate even when there is as much as 25% error in the input data.

4. The calculated Z-component of vectors, such as the Z-component of A_{Rot} and T, is more sensitive to noise than X- and Y-components. The Z-component of the motion is only derivable from the change in the disparities between pairs of images. Since these changes are usually small, noise in the image position will produce inaccurate motions in the Z-direction.

5. Using more input data, e.g., observing an extra frame, decreases the mean and standard deviation of all error.

5.5 Conclusions

We have presented the basic derivation of a method to calculate the general motion parameters when given the locations of a single point in 5 consecutive frames of a motion sequence. This method has been tested on a variety of synthetic and real data. This method has several advantages over most other methods, including a simplier computation method, natural parameters of motion, completely automatic generation of initial guesses, and the ability to handle larger separations between frames. The major weaknesses are in the ability to handle small rotations, leading either to slow convergence or to a description of the motion as a translation. The work presented here is by no means a complete motion system, there still remains much work in incorporating these ideas into a matching system where the predictions of the motion estimation module can aid later matching, and incorrect matches can be detected and eliminated.

Acknowledgements

This research was supported in part by DARPA contracts DACA76-85-C-0009 and F33615-84-K1404, order No. 3119 and monitored by the U.S. Army Engineer Topographic Laboratories and the Air Force Wright Aeronautical Laboratories, respectively.

References

Adiv, G., (1985) 'Determining 3-dimensional motion and structure from optical flow generated by several moving objects,' *IEEE Trans. on Pattern Analysis and Machine Intelligence,* vol. PAMI-7, no. 4, pp. 384-401.

Aggarwal, J.K., and Mitiche, A., (1985) 'Structure and motion from images,' *Proc. of Image Understanding Workshop,* Miami Beach, pp. 89-97.

Ballard, D.H., and Kimball, O.A., (1983) 'Rigid body motion from depth and optical flow,' *Computer Vision, Graphics and Image Processing,* vol. 22, no. 1, pp. 95-115.

Bolles, R.C., and Baker, H.H., (1985) 'Epipolar-plane image analysis: A techinque for analyzing motion sequences,' *Proc. IEEE 3rd Workshop on Computer Vision: Representation and Control,* Bellair, MI, pp. 168-178.

Broida, T., and Chellappa, R., (1985) 'Estimation of object motion parameters from noisy images,' *Proc. IEEE Conf. on Computer Vision and Pattern Recognition,* San Francisco, pp. 82-88.

Broida, T., and Chellappa, R., (1986) 'Kinematics and structure of a rigid object from a sequence of noisy images,' *IEEE Workshop on Motion: Representation and Analysis,* Kiawah Island, SC, pp. 95-100.

Fang, J.Q., and Huang, T.S., (1984) 'Solving 3-dimensional small-rotation motion equations: Uniqueness, algorithms and numerical results,' *Computer Vision, Graphics and Image Processing,* vol. 26, no. 1, pp. 183-206.

Faugeras, O.D., and Price, K.E., (1981) 'Semantic description of aerial images using stochastic labeling,' *IEEE Trans. on Pattern Analysis and Machine Intelligence,* vol. PAMI-3, no. 6, pp. 633-642.

Hoffman, D.D., and Flinchbaugh, B.E., (1982) 'The interpretation of biological motion,' *Biological Cybernetics*, vol. 42, pp. 195-204.

Lappin, J.S., Doner, J.F., and Kottas, B.L., (1980) 'Minimal conditions for the visual detection of structure and motion in three dimensions,' *Science,* vol. 209, pp. 717-719.

Lawton, D.T., (1980) 'Constraint-based interference from from image motion,' *Proc. of the 1st Annual Natl. Conf. on Artificial Intelligence,* Stanford.

Mitiche, A., Seida, S., and Aggarwal, J.K., (1985) 'Determining position and displacement in space from images,' *Proc. of IEEE Conf. on Computer Vision and Pattern Recognition,* San Francisco, pp. 504-509.

Ohlander, R., Price, K., and Reddy, D.R., (1978) 'Picture segmentation using a recursive region splitting method,' *Computer Graphics and Image Processing,* vol. 8, pp. 313-333.

Shariat, H., (1985) 'The motion problem: A decomposition-based solution,' *Proc. of IEEE Conf. on Computer Vision and Pattern Recognition,* San Francisco, pp. 181-183.

Shariat, H., (1986) 'The motion problem: How to use more than two frames,' Dept. of Electrical Engineering, Univ. of Southern California, PhD dissertation.

Tsai, R.Y., and Huang, T.S., (1981) 'Estimating three-dimensional motion parameters of a rigid planar patch,' *Proc. of IEEE Conf. on Pattern Recognition and Image Processing*, Dallas.

Tsai, R.Y., and Huang, T.S., (1984) 'Uniqueness and estimation of three-dimensional motion parameters of rigid objects with curved surfaces,' *IEEE Trans. on Pattern Analysis and Machine Intelligence*, vol. PAMI-6, pp. 13-27.

Ullman, S., (1979) **The Interpretation of Visual Motion**, MIT Press, Cambridge, MA.

Webb, J.A, and Aggarwal, J.K., (1982) 'Structure from motion of rigid and jointed objects,' *Artificial Intelligence*, vol. 19, pp. 107-130.

Yasumoto, Y., and Medioni, G., (1985) 'Experiments in estimation of 3-dimensional motion parameters from a sequence of image frames,' *Proc. of IEEE Conf. on Computer Vision and Pattern Recognition*, San Francisco, pp. 89-94.

Yen, B.L., and Huang, T.S., (1983) 'Determining 3-dimensional motion and structure of a rigid body using the spherical projection,' *Computer Vision, Graphics and Image Processing*, vol. 21, pp. 21-32.

CHAPTER 6

An Experimental Investigation
of Estimation Approaches for Optical Flow Fields

W. Enkelmann
R. Kories
H.-H. Nagel
G. Zimmermann

6.1 Introduction

The halftone image of a scene can be described by giving the gray value as a function of the image plane coordinate vector x, i.e., the picture function $g(x)$. In the case of relative motion between the camera and the depicted scene, the picture function will depend not only on the image plane coordinate vector x, but in addition on the time t. During a short enough time interval, Δt, the resulting temporal change of the picture function can be approximated by a local shift of spatial gray value structures. This apparent shift can be described by an optical flow vector field, $u(x,t)$. It maps the gray value $g(x-u\Delta t,t)$ recorded at time t at the image plane location $x-u\Delta t$ into the gray value $g(x,t + \Delta t)$ recorded at location x at time $t + \Delta t$. In most cases, this optical flow field is a good approximation to the temporal displacement of the image of a depicted surface element between time t and time $t + \Delta t$.

Various approaches have been suggested to estimate the optical flow field from image sequences (see, for example, Nagel, 1983c; Nagel, 1984;

Nagel, 1985; or Hildreth, 1983 for a review of the literature). This chapter discusses two basically different approaches for the estimation of optical flow vector fields in order to clarify their advantages and disadvantages as well as the relations between them. The first approach is based on the extraction and interframe match of features. Several features are investigated using - among others - a large data base consisting of 120 real-world images. The performance of the approach involving features is evaluated regarding its ability to allow the detection of a moving car in a cluttered parking lot. One particular feature extractor, the monotonicity operator, allows a detection rate of about 90% with only few false alarms. The second figure of merit is the estimation uncertainty of the car's displacement. Again, the monotonicity operator performs best in our experiments using different features. Its estimation uncertainty is on the order of half a pixel.

This operator which describes local properties of the gray value function in a heuristic manner is investigated further. Details are treated in Section 6.2. It is found that two components of this operator - associated with peaks and pits of the picture function - provide the essential contributions to the optical flow estimation.

In distinction to feature-based approaches, one may directly evaluate the spatio-temporal variation of the picture function which results in a differential or gradient-based approach. This gradient-based approach has been supplemented by a smoothness requirement on the optical flow field, $u(x)$ (Horn and Schunk, 1981), in which the estimation of $u(x)$ was formulated as an optimization problem resulting in a system of partial differential equations for the optical flow field. Subsequent investigations - recapitulated in Section 6.3 - showed that significant contributions for the determination of $u(x)$ could be expected among others from gray value structures which correspond to the peaks and pits evaluated in the feature-based approach discussed in Section 6.2. Section 6.4 compares results from both approaches obtained using the same input data.

6.2 Feature Based Estimation

6.2.1 The Monotonicity Operator

The basic idea of the monotonicity operator is illustrated using Figure 1. The gray value of the central pixel A is compared to the gray values of

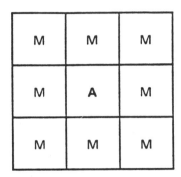

Figure 1. Monotonicity mask.

the adjacent pixels M. The result of the monotonicity transformation for pixel A is obtained according to the following rule:

The central pixel A is classified into one of the classes 0 to 8 according to the number of gray values which are less than the gray value in the center.

A pixel which represents a local maximum belongs to class 8 because all the surrounding pixels have a smaller gray value. Pixels belonging to class 0 are part of a gray level plateau or represent local minima. A pixel that is part of a gray value structure that looks like a roof top is part of class 6 due to the fact that all three pixels on the ridge have the same intensity as the center pixel. Six of them are lower than the central pixel. If the roof top is tilted, the central pixel is assigned to class 7. Obviously, this definition yields for any topological structure exactly one class of the monotonicity transformation. This statement, however, does not remain valid if read in the opposite direction: consider that the class of a pixel is not changed by a permutation of its neighbors. The monotonicity operator offers some advantages:

- The basic operation is very simple. Only gray value comparisons among nearby pixels have to be performed.

- A hardware realization appears feasible.

- A class assignment does not depend on the absolute gray value.

- A class assignment is virtually independent of contrast.

Relevant for the assignment of a pixel to a specific class are not the gray value differences but their rank. This implies:

- The monotonicity operation is invariant under monotone gray value transformations (the attribute giving rise to the name of the operator).

- Each pixel of an image is classified. This is also true for image areas with almost no visible structures and even for regions of constant gray values that usually are not extracted by other feature extractors.

The invariance properties of the monotonicity operator guarantee that the operation is insensitive against changes of the camera gain, against shading inhomogeneities of the sensor as well as against certain types of illumination changes. The operator assigns each pixel to one class. This is a mapping of a gray value image into nine binary images. According to the definition, all the classes are disjoint and thereby form a partition. An important property is that pixels tend to form blobs in the image (see Figure 4 and 5). Blobs consisting of less than ten pixels are eliminated in order to reduce the computational burden. Blobs touching the image borders are deleted as well, since the centroids of such blobs are moving generally slower than the actual velocity of the image area giving rise to such blobs.

The stability of the blobs is increased considerably if high and low frequencies of the image signal are suppressed by a bandpass filter before the application of the monotonicity transform. The reason is that ordinary video signals contain interferences like hum and high frequency noise which are reduced by the filter. The second advantage is that the filter produces more compact blobs which are less likely to split and merge over time. The bandpass filter is realized according to a proposal by Burt (1981), and operates approximately like a Difference-of-Gaussians (DOG) filter. Figure 2 shows the gain plot of the filter.

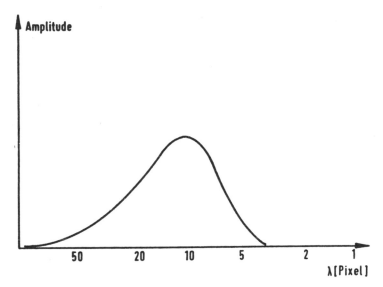

Figure 2. Gain of the bandpass-filter as a function of the wavelength λ.

It appears inappropriate to sample the bandpass filtered image with the mask displayed in Figure 1 in that no prominent signal with a wavelength of less than two pixels is present in the signal after the filtering. For this reason, the sampling scheme has been modified as shown in Figure 3. The central pixel and the sample points are not necessarily adjacent. They are separated by a sampling distance L.

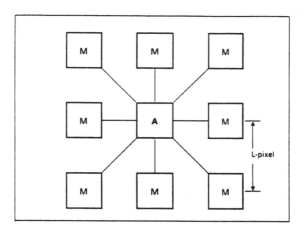

Figure 3. Modified sampling scheme of the monotonicity operator.

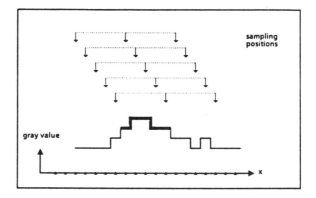

Figure 4. Localization of gray value structure "maximum" illustrated in one dimension. The sampling distance L = 5 is the reason that an extended blob has the property "hill".

The monotonicity operator does not contain explicitly any threshold or other tunable parameter - see, however, the more detailed discussion in Section 6.2.6. The gray value resolution of the sensor defines the minimal separable gray value. If the number of gray levels is appropriate to the signal-to-noise ratio of the image, then so is the monotonicity operator.

6.2.2 From Feature Positions to Optical Flow Vectors

The individual classes associated with the monotonicity operator represent topological structures of the picture function. Regarding the estimation of local displacements within the image, it is useful to determine the localization of structures having the same topological properties and to trace them from image to image. Adjacent pixels of the same class are merged into blobs. This is performed by a component labeling process. The centroid position of a blob, its area, and its class number are used as feature description. This results in a very compact description for each image of the sequence.

Tracing the segmented blobs from image to image allows the estimation of optical flow vector fields. The correspondence problem is simplified by the fact that only blobs belonging to the same class are considered as matching partners. This heuristic is justified by the fact that the topological properties of blobs do not change rapidly. In case there is more than one blob available for matching, the one with the most similar area is chosen. Tracing the centroids of the matched blobs for the duration of two or more images defines a local optical flow vector.

6.2.3 Test Sequence

Figure 5 shows one image of a sequence which consists of 120 video images. The sequence shows a car moving in a parking lot from right to left. The video camera has been manually traversed to the left during the recording of the sequence in order to keep the image of the moving car approximately centered within the image frame.

The velocity of the car relative to the stationary background increases during the sequence. The car is partially occluded during its motion due to parked vehicles. The image of the moving car covers only 1.5 to 2.5 percent of the total image area.

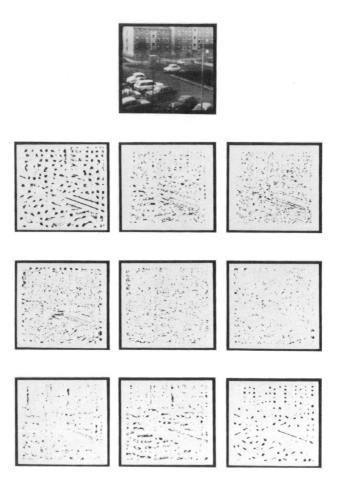

Figure 5. Nine monotonicity classes derived from the original image
on top.

At the beginning of the sequence, the background around the moving
car is well structured showing high contrast areas. The foreground shows
mainly a paved road with uniform gray value. Such areas are difficult to
process by methods that look for prominent image structures.

The scene was recorded on a cloudy day. Nevertheless, the vehicle
still cast a slight shadow that evidently was moving as well. The camera
and vehicle motion produced motion blur, while the metal and glass

surfaces of the vehicles produced highlights with parts of the image being so bright as to saturate the camera.

We consider this test sequence as a very difficult example presenting most of the problems with which a displacement estimation algorithm analyzing natural scenes must deal.

Figure 5 shows an image from the test sequence and the result of applying the monotonicity operator to the given image. Figure 6 displays the result of estimating the local optical flow vectors. The first field, Figure 6b, is derived tracing the features for the duration of two images. The next fields display the flow vectors for subsequences comprising three, four, five, and six image frames. It can be observed that the number of vectors decreases as the number of images taken into account increases. The number of vectors which do not correctly describe the proper displacement, however, decreases much more strongly. The algorithm requires that a feature appear repeatedly in all the images considered for tracing. This approach drops unstable features and therefore results in more reliable vector fields.

6.2.4 Moving Object Detection

Visual examination of the displacement vector fields displayed in Figure 6 suggests that the car can be detected already due to the difference between its optical flow vectors and the flow vectors of the background. A local clustering of vectors similar to region growing approaches for image segmentation allows a separation of image areas with different optical flow vectors. Based upon this separation, the automatic detection of objects moving relative to their environment becomes feasible. The frames in Figure 7 indicate the result of our detection algorithm. Frequently the moving vehicle can be detected. However, sometimes areas that are not actually moving relative to the stationary background are cued by the algorithm. The cause is that local features have changed considerably and do not produce the correct optical flow vectors. If such a problem occurs for

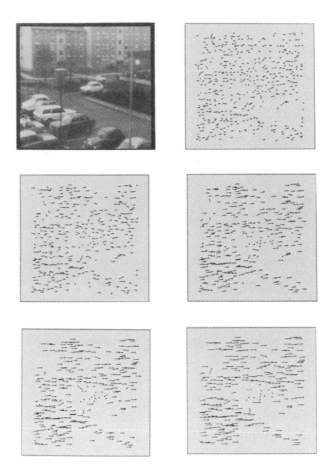

Figure 6. Optical flow vector fields obtained from the test sequence for
(b) 2, (c) 3, (d) 4, (e) 5, (f) 6 consecutive images starting with the image shown on top.

adjacent vectors they may form a cluster that gives rise to the interpretation
as a moving object. Such an event is denoted as a false alarm. The most
severe case we ever observed in our investigation is displayed in Figure 7b
Three false alarms were produced. The examples in the bottom row of
Figure 7 show almost perfect results.

Figure 7. Typical detection results.

6.2.5 Performance Analysis of the Monotonicity Operator

We analyzed the performance of the optical flow estimation algorithm based on features produced by the monotonicity operator according to different figures of merit. The first figure of merit is the number of detections of the moving car that

can be achieved evaluating the vector fields. The accuracy of the displacement estimation for a moving object is used as a second performance measure.

Our investigations included feature extractors which have been used elsewhere for processing time-varying imagery. These feature extractors are:

-   A point detector (Moravec, 1977) which calculates the sum of squared differences of adjacent pixels in the horizontal, vertical, and the two diagonal directions. Points with a local maximum for the smallest value of these four sums are used as characteristic points.

-   A different point detector (Dreschler and Nagel, 1983) which detects points with particular curvature combinations of the picture function in a 5 × 5 neighborhood. This combination of curvatures is related to "gray value corners" (Nagel, 1983d).

-   A line finder (Korn and Kories, 1980) which uses a one-dimensional bandpass filter similar to the "mexican hat" operator with standard deviation of 1.5 to yield edge points that are merged into straight line segments resulting in an approximation of object boundaries.

For all four operators, the positions of the features have been fed into the optical flow estimation algorithm. Every one of the 120 images of the test sequence has been processed by each of these four feature extractors. Tracking the features in the images for five consecutive images resulted in 116 optical flow vector fields for each operator. All of these optical flow fields have been submitted to the detection algorithm. The frames, cueing moving objects, have been inserted into the appropriate original images. Subsequently, it has been verified visually whether they covered the moving car. If so, they are counted as proper detections. Those not covering the vehicle were considered false alarms. All the vector fields for each operator have been processed with the same parameters.

Detection Results

Figure 8 illustrates the performance of the various feature detectors.

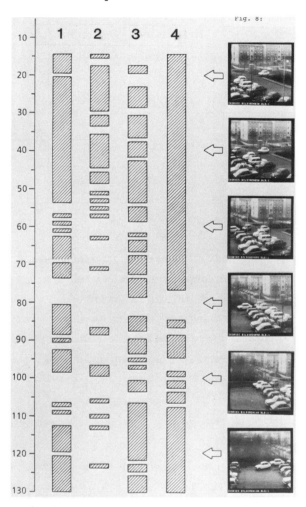

Figure 8. Detection results based on the following features: 1 Straight lines, 2 Moravec, 3 Combination of picture function curvatures, 4 Monotonicity-Operator.

The hatched bars indicate where the algorithm succeeded in detecting the car. Those images of the sequence where the vehicle is occluded (image

70 to 105) posed a problem to all the feature extractors. An evaluation
results in the following:

Operator	Detections	Detection-rate in %	False alarms
Straight lines	86	74 ± 4.7	76
Moravec	41	35 ± 7.4	53
Curvature combination	84	72 ± 4.9	107
Monotonicity	103	89 ± 3.1	53

False alarms and detections may occur simultaneously in the same
image. The sum of both may, therefore, exceed the total number of
images. Obviously, among the feature extraction operators considered, the
monotonicity operator performs best for this sequence. The straight line
finder operator and the detector based on a combination of gray value curvatures produce similar detection rates. They differ, however, considerably
in the number of false alarms. Therefore, a more detailed consideration is
required.

The detection algorithm may commit two types of errors. It may
miss the moving car or it may cue a stationary area as moving, i.e., a false
alarm. Depending on the actual application of the detection algorithm, the
false alarm rate may be rated differently. If the method is used to direct
the attention of an observer towards moving objects, one may tolerate certain false alarms. A much lower false alarm rate will be required if a false
alarm during the surveillance of an industrial robot for collision avoidance
results in emergency stops. For a particular application the two types of
errors will be associated with a certain cost. We define a cost function that

depends on the number of false alarms as well as on the number of misses. It is scaled such that it lies between zero and one. A given detection algorithm will be considered the most appropriate if it minimizes the total cost as defined by the following:

$$c_{tot} = c_f F + (1-c_f)M. \tag{1}$$

F represents the number of false alarms and M the number of misses. The total cost, c_{tot}, versus the relative false alarm cost, c_f, is plotted for each operator in Figure 9.

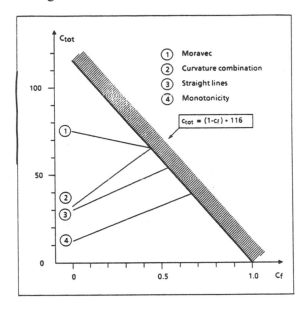

Figure 9. Cost function for the different operators.

For all values of the relative false alarm cost the monotonicity operator produces the lowest total cost. The straight line finder and the gray value corner detector behave similar to one another, while the behavior of the Moravec operator is totally different. The Moravec operator frequently is not able to detect the vehicle but produces only a false alarm rate that is equal to that of the monotonicity operator. The total cost for the Moravec operator, therefore, decreases with increasing relative false alarm cost.

Assume that the density of the vectors in an optical flow vector field is decreased continually, then the detection rate would be reduced as well as the number of false alarms. The cost plot in Figure 9 would rise on the left side and sink on the right. In the limiting case F and M go to zero as symbolized by the hatched area in which it is inappropriate to perform a detection algorithm. The cost would be higher than not using the motion detector at all.

Accuracy of Object Velocity Estimation

For the detection of moving objects, the displacement of different image areas is determined. Figure 10 shows the velocity of the background in the upper curve. The lower curve is that of the car's horizontal velocity component measured by the monotonicity operator in those images where the vehicle had been detected automatically. The irregular motion of the camera due to the manual operation of the TV-camera can be recognized. The velocity of the car relative to the background is shown in Figure 11. The car accelerates during the sequence. The ordinate shows negative values because the motion occurs against the direction of the image coordinates. The velocity curve is quite smooth and unaffected by the rough camera motion.

In order to derive an estimation of the accuracy of the velocity measurement, we established a reference measurement of the car's projected velocity. It was traced manually within the image sequence by locating five to fifteen different features like headlights, window corners, etc. on the car. The mean displacement vector for these handpicked features was considered to be the true displacement. Figure 12 shows this reference velocity of the car by the continuous curve, while the dots represent measurements of the monotonicity operator whenever it was able to detect the car correctly. A detailed error analysis comparing all the measurements of the operators to each other yields the following accuracies (see Table 1).

Operator	Accuracy
visual reference	0.54 ± 0.07 pixel
Straight lines	0.46 ± 0.07 pixel
Moravec	0.66 ± 0.14 pixel
Curvature combination	0.68 ± 0.10 pixel
Monotonicity	0.45 ± 0.06 pixel

Table 1. Accuracies of the displacement estimation.

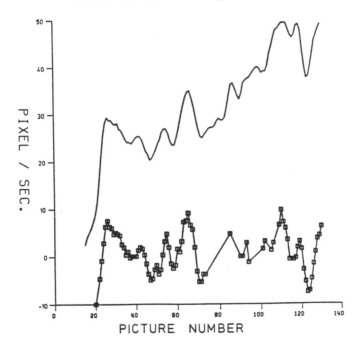

Figure 10. Velocity component in horizontal direction for the background (upper curve) and the vehicle (lower curve).

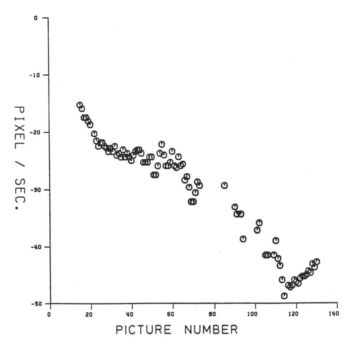

Figure 11. Velocity component of the vehicle in horizontal direction
relative to the background.

6.2.6 Robustness of the Monotonicity Operator Against Parameter Changes

The monotonicity operator has basically two parameters. One is the
center wavelength of the bandpass filter, the second is the sampling distance of the operator mask. We suspected that the parameters should be
matched to each other in order to produce stable features. We defined a
stability coefficient which is the percentage of the features that survive for
five images, i.e., that produce vectors. In an investigation, we combined
several bandpass filters (see Figure 13) with monotonicity masks of sampling distances 3, 5, and 7. For each combination we computed optical

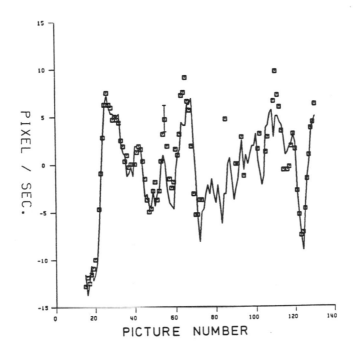

Figure 12. Reference horizontal velocity component of the vehicle (curve) and measurements of the monotonicity operator (dots).

flow vector fields and determined their stability coefficients. They are (in percent):

Sampling distance

		3	5	7
Center frequency	5	16	30	29
of	10	25	36	32
bandpass filter	20	12	24	24

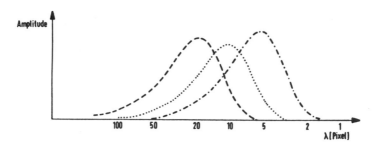

Figure 13. Gain curves of the band-pass filters.

It can be seen that matching of center frequency and sampling distance is not very critical. Low frequencies and small sampling distances, however, are not a good combination, as can be seen by inspection of the vector fields. The best result was achieved with the vector field displayed in the center of Figure 14. More than one third of the extracted blobs can be traced for the duration of five images. The total number of features associated with the car - which is crucial for the moving object detection - is especially large in this frequency band. Obviously, the bandpass filter is appropriate for the details of this vehicle.

The results indicate that even a mismatch of center frequency and sampling distance does not totally deteriorate the algorithm. On the other hand, the stability coefficient may serve as an excellent indicator of the desirability of switching to a different channel during the processing of an image sequence.

6.2.7 Reduction to Two Classes

The shapes of the blobs in distinct monotonicity classes are very different as can be seen in Figure 5. We, therefore, suspected that some classes may be more stable than others and consequently more suitable for tracking. An appropriate figure of merit for the stability of the blobs is the stability coefficient. The upper half of Figure 15 shows full bars indicating the number of blobs that could be found in the different classes processing

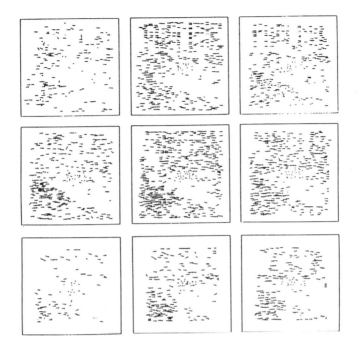

Figure 14. Vector fields obtained by combinations of band-pass filters (vertical) and sampling distances (horizontal).

a particular sequence of 50 images. The vertical bars indicate the minimal and maximal number in one image. The lower part of Figure 15 describes by dotted bars the number of vectors computed from tracking the blobs. It can be seen that all the classes contribute approximately the same number of vectors. However, classes zero and eight produce the most stable features, i.e., a higher ratio of features survive in all images. In order to reduce the computational effort, we decided to use class zero and eight as the main features for optical flow estimation. In order to obtain symmetric behavior of these two classes, we changed the definition of class zero: eight neighbors are required to have a gray value higher than the center pixel. This definition implies a very simple topological interpretation. Features from class zero can be interpreted as pits and features from class eight as hill tops.

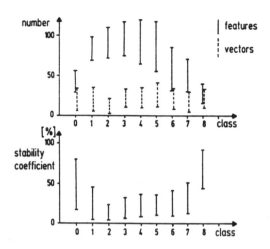

Figure 15. Upper plot: Number of features and vectors in different monotonicity classes. Lower plot: Stability coefficient.

Experiments confirmed that these two features are most suited for displacement estimation. This fact is already well known from theoretical considerations for displacement estimation based on analytical modelling of the picture function (Nagel, 1984). The next section describes a method that systematically derives optical flow vector estimates from local properties of the picture function.

6.3 Analytical Approach for the Estimation of Optical Flow Vector Fields

In most situations, the mere gray value variations do not provide sufficient information to completely determine the planar optical flow vector field, $u(x) = (u(x),v(x))^T$, which links the pixel position, $x = (x, y)^T$, in one frame, g1, to the corresponding position in another frame, g2. Therefore, it has been postulated that an optical flow vector field should vary smoothly as a function of the image coordinate vector, x (Horn and Schunck, 1981). This postulate enabled them to estimate both components, u and v, of the optical flow vector, u(x). Such a general smoothness requirement, however, forces the estimated optical flow vector field to vary smoothly even across the image of occluding edges. To cope with this difficulty, an "oriented smoothness" requirement has been introduced which constrains the variation of the optical flow vector field only in those directions along which an optical flow vector component can not be inferred from spatio-temporal gray value changes (Nagel, 1983a and 1983b).

6.3.1 The "Oriented Smoothness" Constraint

The original approach (Horn and Schunck, 1981) formulated the estimation of an optical flow vector field as a minimization problem. That approach has been adapted (Nagel, 1983b) as follows:

$$\iint \left\{ \left[g2(\mathbf{x}) - g1(\mathbf{x} - \mathbf{u}) \right]^2 + a^2 trace\left[(\nabla\mathbf{u})^T \mathbf{W}(\nabla\mathbf{u}) \right] \right\} dxdy \qquad (2)$$

$$\Rightarrow min.$$

The local gray value variation influences the oriented smoothness constraint directly through the weight matrix, $\mathbf{W}$, which is defined as

$$\mathbf{W} = \frac{\mathbf{F} + \gamma\mathbf{I}}{trace(\mathbf{F} + \gamma\mathbf{I})} \qquad (3)$$

with

$$\mathbf{F} = \left\{ \begin{bmatrix} g_y \\ -g_x \end{bmatrix} \begin{bmatrix} g_y \\ -g_x \end{bmatrix}^T + b^2 \begin{bmatrix} g_{yy} & -g_{xy} \\ -g_{xy} & g_{xx} \end{bmatrix} \begin{bmatrix} g_{yy} & -g_{xy} \\ -g_{xy} & g_{xx} \end{bmatrix}^T \right\}. \qquad (4)$$

Subscripts x and y denote partial differentiation with respect to the corresponding image coordinate variables. As discussed in detail by

Enkelmann (1985) this formulation of an oriented smoothness constraint avoids problems which occur in image regions with an almost homogeneous gray value distribution or with constant gray values along one direction, i.e., a straight line gray value transition. The estimation of optical flow vector fields using weight matrices such as $\mathbf{W}=\mathbf{F}$ or $\mathbf{W}=\mathbf{F}/\text{trace}(\mathbf{F})$ have been investigated by Nagel and Enkelmann (1986).

The Euler-Lagrange equations for the minimization problem in Eq. 2 yield

$$-\left[g2(\mathbf{x}) - g1(\mathbf{x}-\mathbf{u})\right] \frac{\partial g1(\mathbf{x}-\mathbf{u})}{\partial u} \tag{5a}$$

$$- a^2 \begin{bmatrix} d/dx \\ d/dy \end{bmatrix}^T \frac{\mathbf{F}+\gamma\mathbf{I}}{\text{trace}(\mathbf{F}+\gamma\mathbf{I})} \begin{bmatrix} u_x \\ u_y \end{bmatrix} = 0$$

and

$$-\left[g2(\mathbf{x}) - g1(\mathbf{x}-\mathbf{u})\right] \frac{\partial g1(\mathbf{x}-\mathbf{u})}{\partial v} \tag{5b}$$

$$- a^2 \begin{bmatrix} d/dx \\ d/dy \end{bmatrix}^T \frac{\mathbf{F}+\gamma\mathbf{I}}{\text{trace}(\mathbf{F}+\gamma\mathbf{I})} \begin{bmatrix} v_x \\ v_y \end{bmatrix} = 0.$$

In order to simplify the subsequent attempts to solve this system of coupled partial differential equations, the components of the weight matrix $\mathbf{W}$ are considered to vary slowly from pixel to pixel so that derivatives of $\mathbf{W}$ can be neglected. This assumption enables the simplification of Eqs. 5a and b into the following:

$$-\left[g2(\mathbf{x}) - g1(\mathbf{x}-\mathbf{u})\right] \frac{\partial g1(\mathbf{x}-\mathbf{u})}{\partial u} \tag{6a}$$

$$- a^2 \, trace\left[\frac{F+\gamma I}{trace(F+\gamma I)} \begin{bmatrix} u_{xx} & u_{xy} \\ u_{xy} & u_{yy} \end{bmatrix}\right] = 0$$

and

$$-\left[g2(\mathbf{x}) - g1(\mathbf{x}-\mathbf{u})\right] \frac{\partial g1(\mathbf{x}-\mathbf{u})}{\partial v} \tag{6b}$$

$$- a^2 \, trace\left[\frac{F+\gamma I}{trace(F+\gamma I)} \begin{bmatrix} v_{xx} & v_{xy} \\ v_{xy} & v_{yy} \end{bmatrix}\right] = 0.$$

It is assumed that an approximate solution, $u_0(\mathbf{x})$, for the displacement vector field is known. The true solution, $u(\mathbf{x})$, should differ from $u_0(\mathbf{x})$ by a small correction vector field, $Du(\mathbf{x})$, as indicated in the following:

$$u(\mathbf{x}) = u_0(\mathbf{x}) + Du(\mathbf{x}). \tag{7}$$

In order to simplify the notation, the argument $\mathbf{x}$ for the approximate displacement vector field $u_0(\mathbf{x})$, as well as for the correction vector field, $Du(\mathbf{x})$, will be omitted.

Since Du is considered to be small, we retain only first order terms in the components of Du:

$$g1(\mathbf{x}-\mathbf{u}) = g1(\mathbf{x}-u_0-Du) \simeq g1(\mathbf{x}-u_0) - (\nabla g1(\mathbf{x}-u_0))^T Du \tag{8}$$

with

$$\frac{\partial g1(\mathbf{x}-\mathbf{u})}{\partial u} \simeq \frac{\partial g1(\mathbf{x}-\mathbf{u}_0)}{\partial u} = -\frac{\partial g1(\mathbf{x}-\mathbf{u}_0)}{\partial x} \tag{9a}$$

and

$$\frac{\partial g1(\mathbf{x}-\mathbf{u})}{\partial v} \simeq \frac{\partial g1(\mathbf{x}-\mathbf{u}_0)}{\partial v} = -\frac{\partial g1(\mathbf{x}-\mathbf{u}_0)}{\partial y}. \tag{9b}$$

The effect of a correction, Du, on the terms containing $\nabla\nabla u$ and $\nabla\nabla v$ in Eqs. 6a and b is supposed to contribute only to the central pixel of the operator masks for the calculation of the partial derivatives. The substitution of $\mathbf{u}_0 + \mathbf{Du}$ for $\mathbf{u}$ at the center pixel has the following effect:

$$trace\left[\frac{\mathbf{F}+\gamma\mathbf{I}}{trace(\mathbf{F}+\gamma\mathbf{I})}\begin{bmatrix} u_{xx} & u_{xy} \\ u_{xy} & u_{yy} \end{bmatrix}\right] \simeq \tag{10a}$$

$$trace\left[\frac{\mathbf{F}+\gamma\mathbf{I}}{trace(\mathbf{F}+\gamma\mathbf{I})}\begin{bmatrix} u_{0_{xx}} & u_{0_{xy}} \\ u_{0_{xy}} & u_{0_{yy}} \end{bmatrix}\right] - mDu$$

and

$$trace\left[\frac{\mathbf{F}+\gamma\mathbf{I}}{trace(\mathbf{F}+\gamma\mathbf{I})}\begin{bmatrix} v_{xx} & v_{xy} \\ v_{xy} & v_{yy} \end{bmatrix}\right] \simeq \tag{10b}$$

$$trace\left[\frac{\mathbf{F}+\gamma\mathbf{I}}{trace(\mathbf{F}+\gamma\mathbf{I})}\begin{bmatrix} v_{0_{xx}} & v_{0_{xy}} \\ v_{0_{xy}} & v_{0_{yy}} \end{bmatrix}\right] - mDv.$$

Introduction of these approximations into Eqs. 6a and b yields a system of equations which are linear in the components of the correction vector, Du, as follows:

$$- \left[g2(\mathbf{x}) - g1(\mathbf{x}-\mathbf{u}_0) + (\nabla g1(\mathbf{x}-\mathbf{u}_0))^T \mathbf{Du} \right] \frac{\partial g1(\mathbf{x}-\mathbf{u}_0)}{\partial x} \qquad (11a)$$

$$+ a^2 trace \left[\frac{\mathbf{F}+\gamma\mathbf{I}}{trace(\mathbf{F}+\gamma\mathbf{I})} \nabla\nabla u_0 \right] - m\mathbf{Du} = 0$$

and

$$- \left[g2(\mathbf{x}) - g1(\mathbf{x}-\mathbf{u}_0) + (\nabla g1(\mathbf{x}-\mathbf{u}_0))^T \mathbf{Du} \right] \frac{\partial g1(\mathbf{x}-\mathbf{u}_0)}{\partial y} \qquad (11b)$$

$$+ a^2 trace \left[\frac{\mathbf{F}+\gamma\mathbf{I}}{trace(\mathbf{F}+\gamma\mathbf{I})} \nabla\nabla v_0 \right] - m\mathbf{Dv} = 0.$$

This can be transformed into the following matrix equation:

$$\left[(\nabla g1)(\nabla g1)^T + a^2 m\mathbf{I} \right] \mathbf{Du} = - \left[g2(\mathbf{x})-g1(\mathbf{x}-\mathbf{u}_0) \right] \nabla g1 \qquad (12)$$

$$+ \frac{a^2}{trace(\mathbf{F}+\gamma\mathbf{I})} \left[\begin{array}{c} trace\left[(\mathbf{F}+\gamma\mathbf{I})\nabla\nabla u_0 \right] \\ trace\left[(\mathbf{F}+\gamma\mathbf{I})\nabla\nabla v_0 \right] \end{array} \right]$$

Since the derivatives of g1, as well as of u and v, have to be determined eventually by applying a digital operator to the image area around the position $\mathbf{x}-\mathbf{u}_0$, it appears plausible to average the gradients, $\nabla g1$, over that area. Using this averaging process, Eq. 12 can be written as

$$\left[(\nabla g1)(\nabla g1)^T + \overline{\xi^2}(\nabla\nabla g1)(\nabla\nabla g1)^T + a^2 m\mathbf{I}\right]\mathbf{Du} \tag{13}$$

$$= -\nabla g1\overline{\left[g2(\mathbf{x})-g1(\mathbf{x}-\mathbf{u}_0)\right]} + \frac{a^2}{trace(\mathbf{F}+\gamma\mathbf{I})}\left[\begin{array}{c} trace\left[(\mathbf{F}+\gamma\mathbf{I})\nabla\nabla u_0\right] \\ trace\left[(\mathbf{F}+\gamma\mathbf{I})\nabla\nabla v_0\right] \end{array}\right]$$

where the overbar indicates the averaging over all positions, ξ, in the area around $\mathbf{x}-\mathbf{u}_0$.

6.3.2 Evaluation at Local Extrema of the Picture Function

It is interesting to study the results of this approach at image locations corresponding to the center position of those blobs of the monotonicity operator which belong to classes 0 and 8. Such image locations are characterized as extrema of the picture function. We may assume that the local coordinate system is aligned with the principal curvature directions of the gray value distribution within the operator window, i.e., $g1_{xy}=0$. Let us assume that the approximate solution of the displacement vector field $u_0(\mathbf{x})=0$ implies $\nabla\nabla u_0=\nabla\nabla v_0=0$, the second term on the right hand side of Eq. 13 vanishes.

We may now specialize Eq. 13 to the case of a gray value extremum in frame 1 at location, $\mathbf{x}-\mathbf{u}_0$. A gray value extremum can be characterized by the following equations:

$$g_x = \frac{\partial g}{\partial x} = 0, \qquad g_{xx} = \frac{\partial^2 g}{\partial x^2} \neq 0, \tag{14a}$$

$$g_y = \frac{\partial g}{\partial y} = 0, \quad \text{and} \quad g_{yy} = \frac{\partial^2 g}{\partial y^2} \neq 0. \tag{14b}$$

Insertion of these values into Eq. 13 yields

$$
\begin{bmatrix} \overline{\xi^2} g1_{xx}^2 + a^2 m & 0 \\ 0 & \overline{\xi^2} g1_{yy}^2 + a^2 m \end{bmatrix} Du = \tag{15}
$$

$$
\overline{-\nabla g1 \left[g2(\mathbf{x}) - g1(\mathbf{x}-\mathbf{u}_0) \right]} = -\overline{\xi^2} \begin{bmatrix} g2_x \cdot g1_{xx} \\ g2_y \cdot g1_{yy} \end{bmatrix},
$$

or for the components of the correction vector, Du,

$$
Du = - \frac{g2_x}{g1_{xx}} \frac{1}{1 + \dfrac{a^2 m}{\overline{\xi^2} g1_{xx}^2}}, \tag{16a}
$$

and

$$
Dv = - \frac{g2_y}{g1_{yy}} \frac{1}{1 + \dfrac{a^2 m}{\overline{\xi^2} g1_{yy}^2}}. \tag{16b}
$$

The expressions in Eq. 16 show in a quantitative manner how the first and second partial derivatives of the picture functions, g1(x) and g2(x), influence the estimate of the correction vector, Du, and thus the resulting estimate for u(x) at gray value extrema.

6.4 Discussion

The approaches treated in the previous sections allow the estimation of both components of the optical flow vector at certain image locations, including those which can be characterized as extrema of the picture function (Nagel, 1984). Let us now consider both approaches using the same image sequence. Figure 16 shows the first half-frame of an image sequence recorded from our computer center. The optical flow field for the

Figure 16. First half-frame of image sequence QUE.

first five half-frames of the image sequence obtained with a multigrid algorithm (see Enkelmann, 1985 and 1986) is shown in Figure 17.

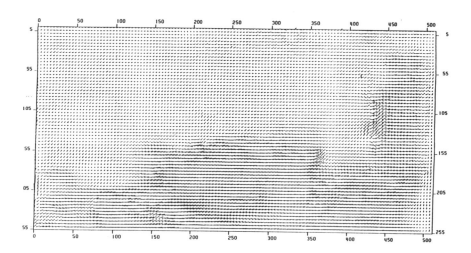

Figure 17. Optical flow field for the first five half-frames of image sequence QUE estimated with the multigrid approach. A subsampling has
been done with a sampling distance of four.

The multigrid algorithm uses the solution approach of Eq. 13 within a
hierarchy of grid levels for the image data as well as for the optical flow
field. One cycle between the coarsest and the finest grid level is called a
multigrid cycle. Figure 18 shows the optical flow field for the same

images but estimated with the approach based on the monotonicity operator. The results shown in Figure 19 have been calculated from the optical
flow field in Figure 17. Shown, however, are only optical flow vectors at
those image locations where a feature point has been selected by the monotonicity operator approach. A bilinear interpolation has been used to
specify optical flow vectors at image locations which do not coincide with
grid points. The differences between the vector fields obtained by the
monotonicity operator and by the analytical approach are shown in Figure
20.

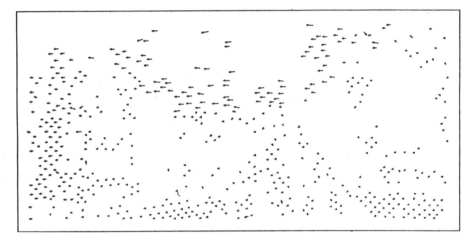

Figure 18. Optical flow field for the first five half-frames of image sequence QUE estimated with the approach based on the monotonicity
operator.

The optical flow vector field for the image section around the disk,
Figure 21, is shown in Figure 22. In order to get a better visual impression of the optical flow field, a subsampling has been performed with a
sampling distance of six. This optical flow field was calculated by con-
catenating the results of the first five half-frames of the image sequence.
The multigrid algorithm for one image pair was stopped after 10 multigrid
cycles across four grid levels each, i.e., after 40 iteration steps. For the
first image pair of the sequence, a start vector field, $u_0(x){=}0$ was used.

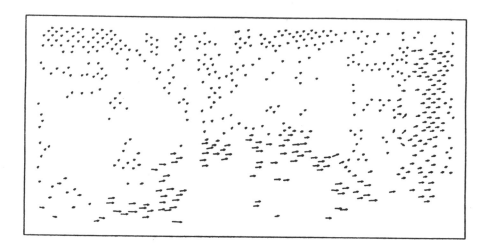

Figure 19. Optical flow field for the first five half-frames of image sequence QUE estimated with the analytical approach, but only at those image locations where a feature point has been selected by the monotonicity operator approach.

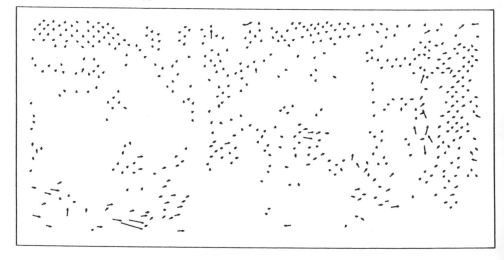

Figure 20. Differences between the results shown in Figure 18 and 19. All difference vectors are scaled by a factor 2.

The result for the first image pair was used as the starting value for the second image pair of the sequence.

In order to show that the multigrid algorithm yields acceptable optical flow vectors in most parts of the image even after ten multigrid cycles, the difference vector field for the results obtained after termination of the multigrid algorithm and after the 10th multigrid cycle is shown in Figure 23. The multigrid algorithm terminated because the maximum of all correction vectors had become smaller than 0.2 units of the image raster. It can be seen that only at the image border, inside the disk, and, partially, in areas with depth discontinuities, the correction vector had not yet dropped below the termination threshold after the 10th multigrid cycle.

The encouraging point about these investigations is the observation that the heuristic estimation approach treated in Section 6.2 and the analytic approach treated in Section 6.3 converge to the same conclusion: certain gray value structures such as extrema of the picture function appear to provide dominant contributions to the estimated optical flow vectors, and the results essentially agree quantitatively within the estimation accuracies. It thus appears worthwhile to study further the analytical approach, as well as its associated computational approaches, and to look for additional situations in which heuristic approaches can be related quantitatively to this emerging theory.

Figure 21. Section of the image shown in Figure 16 selected around the disk.

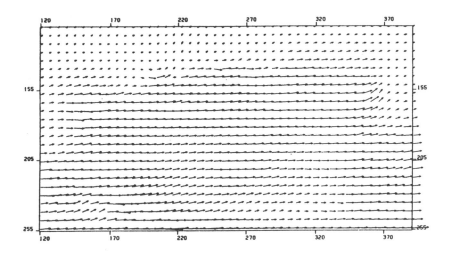

Figure 22. Optical flow field for the image section shown in Figure 21, estimated with the analytical approach. A subsampling with a distance of six has been done in order to enable a better visual impression.

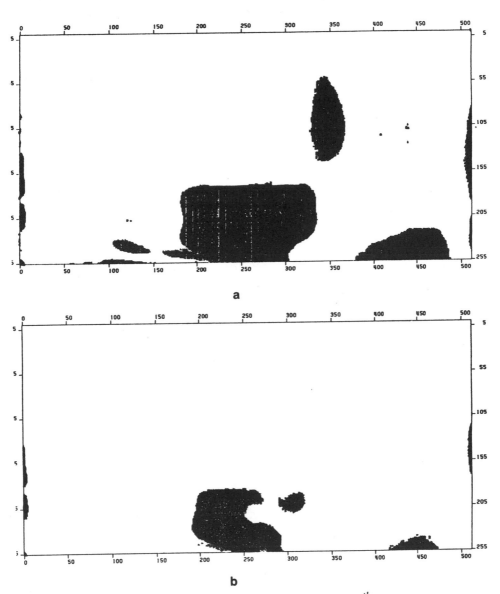

Figure 23. Differences between the vector fields for the 5th pair of half-frames after multigrid cycle 20 and multigrid cycle 10: (a) differences greater than 0.2 units; (b) differences greater than 0.5 units. The multigrid algorithm terminated after multigrid cycle 20 because the maximal magnitude of all correction vectors had become smaller than 0.2 units of the image raster.

References

Burt, P., (1981) 'Fast filter transforms for image processing,' *Computer Graphics and Image Processing,* vol. 16, pp. 20-51.

Dreschler, L.S., and Nagel, H.-H., (1983) 'On the selection of critical points and local curvature extrema of region boundaries for interframe matching,' *Proc. Nato Advanced Study Institute on Image Sequence Processing and Dynamic Scene Analysis,* Braunlage, FR Germany, 1982, T.S. Huang (ed.), Springer-Verlag, Berlin, pp. 457-470.

Enkelmann, W., (1985) 'Mehrgitterverfahren zur Ermittlung von Verschiebungsvektorfeldern in Bildfolgen,' Fachbereich Informatik, Universität Hamburg, Dissertation.

Enkelmann, W., (1986) 'Investigations of multigrid algorithms for the estimation of optical flow fields in image sequences,' *Proc. IEEE Workshop on Motion: Representation and Analysis,* Charleston, SC, pp. 81-87.

Hildreth, E.C., (1983) 'The measurement of visual motion,' Department of Electrical Engineering and Computer Science, MIT, Cambridge, MA, PhD. Dissertation.

Horn, B.K.P., and Schunck, B.G., (1981) 'Determining optical flow,' *Artificial Intelligence,* vol. 17, pp. 185-203.

Korn, A., and Kories, R., (1980) 'Motion analysis in natural scenes picked up by a moving optical sensor,' *Proc. Int. Conf. on Pattern Recognition,* Miami Beach, FL, pp. 1251-1254.

Moravec, H.P., (1977) 'Towards automatic visual obstacle avoidance,' *Proc. Int. Joint Conf. on Artificial Intelligence,* Cambridge, MA, p. 584.

Nagel, H.-H., (1983a) 'On the estimation of dense displacement vector fields from image sequences,' *ACM Interdisciplinary Workshop on Motion: Representation and Perception*, Toronto, Canada, pp. 59-65.

Nagel, H.-H., (1983b) 'Constraints for the estimation of displacement vector fields from image sequences,' *Proc. Int. Joint Conf. on Artificial Intelligence*, Karlsruhe, FR Germany, pp. 945-951.

Nagel, H.-H., (1983c) 'Overview on image sequence analysis,' *Proc. Nato Advanced Study Institute on Image Sequence Processing and Dynamic Scene Analysis*, Braunlage, FR Germany, 1982, in T.S. Huang (ed.), Springer-Verlag, Berlin, pp. 2-39.

Nagel, H.-H., (1983d) 'Displacement vectors derived from second order intensity variations in image sequences,' *Computer Vision, Graphics and Image Processing*, vol. 21, pp. 85-117.

Nagel, H.-H. (1984) 'Recent advances in image sequence analysis,' *Proc. Premier Colloque Image - Traitement, Synthese, Technologie et Applications*, Biarritz, France, pp. 545-558.

Nagel, H.-H., (1985) 'Analyse und Interpretation von Bildfolgen,' *Informatik-Spektrum*, vol. 8, Teil I: pp. 178-200; Teil II: pp. 312-327.

Nagel, H.-H., and Enkelmann, W., (1984a) 'Towards the estimation of displacement vector fields by 'oriented smoothness' constraint,' *Proc. Int. Joint Conf. on Pattern Recognition*, Montreal, Canada, pp. 6-8.

Nagel, H.-H., and Enkelmann, W., (1984b) 'Berechnung von Verschiebungsvektorfeldern in Bildbereichen mit linienhaften oder partiell homogenen Grauwertverteilungen,' *DAGM/ÖAGM Symposium*, Graz, 1984, in W. Kropatsch (ed.), Mustererkennung 1984, Informatik Fachberichte 87, Springer-Verlag, Berlin, pp. 154-160.

Nagel, H.-H., and Enkelmann, W., (1986) 'An investigation of smoothness constraints for the estimation of displacement vector fields from image sequences,' *IEEE Trans. Pattern Analysis and Machine Intelligence,* vol. PAMI-8, no. 5, pp. 565-593.

CHAPTER 7

The Incremental Rigidity Scheme
and Long-Range Motion Correspondence

Shimon Ullman

7.1 The Rigidity-Based Recovery of Structure from Motion

7.1.1 The Perception of Structure from Motion by Human Observers

The human visual system is capable of extracting three-dimensional
shape information from two-dimensional transformations in the image.
Experiments employing shadow projections of moving objects and computer generated displays have established that the three-dimensional shape
of objects in motion can be perceived when their changing projection is
observed, even when each static view is completely devoid of three-
dimensional information.

Observations related to this intriguing capacity were reported as early
as 1860 by Sinsteden (described in Miles, 1931), who first observed the
perception of depth and depth reversals induced by a distant windmill, and
then examined this phenomenon in the laboratory using rotating cardboard
objects. The early experiments in this area were concerned primarily with
the perceived depth reversals of rotating objects. The fact that the three-
dimensional structure of moving objects can be recovered perceptually
from their changing projection was noted by Musatti in 1924 (see Johansson, 1978). It was investigated systematically for the first time using

shadow projections by Wallach and O'Connell (1953) who coined the term, "kinetic depth effect", to describe the perception of three-dimensional structure from motion information. The perception of structure from motion has been investigated extensively since under various conditions, including the motion of connected and unconnected elements, and using both perspective and orthographic projections. For detailed reviews of the extensive research in this area, see (Braunstein, 1976; Johansson, 1978; Ullman, 1979).

7.1.2 Computational Studies of the Recovery of Structure from Motion

In trying to recover three-dimensional structure from the transformations in the image, one is faced with the problem of inherent ambiguity: unless some constraints are imposed, the image transformations are insufficient to determine the three-dimensional structure uniquely. This ambiguity problem has been the focus of a number of computational studies that attempted to discover the conditions under which three-dimensional structure is uniquely determined by the projected transformations in the image.

From the earliest empirical studies of the kinetic depth effect it has been suggested that the rigidity of objects may play a key role in the perception of structure from motion (Wallach and O'Connell, 1953; Gibson and Gibson, 1957; Green, 1961; Johansson, 1975). Computational studies have later established that rigidity is a sufficiently powerful constraint for imposing uniqueness upon the three-dimensional interpretation of the viewed transformations. Uniqueness is defined up to an overall scaling factor for perspective projections, and up to overall distance and a mirror-reflection about the image plane for orthographic projections. A two-dimensional transformation can be tested to determine whether or not it is compatible with the projection of a rigid object in motion. If it is, then the inducing object is in general guaranteed to be unique, and its three-dimensional structure can be recovered. Under orthographic projection the structure is determined uniquely up to a reflection about the image plane. This is an unavoidable ambiguity, since the orthographic projections of a rotating object, and its mirror image rotating in the opposite direction, coincide.

Uniqueness results have been established under a number of different conditions. One of the first proofs showed that under orthographic projection three views of four non-coplanar points are sufficient to guarantee a unique three-dimensional solution (Ullman, 1979a). It has been proved that the instantaneous velocity field and its first and second spatial derivatives at a point admit at most three different three-dimensional interpretations (Longuet-Higgins and Prazdny, 1980). Recently in an elegant analysis it has been shown that, with the exception of a few special configurations, two perspective views of seven points are also sufficient to guarantee uniqueness (Tsai and Huang, 1982). Additional uniqueness results have been obtained for situations where certain restrictions are imposed upon the objects or their motion, such as planar surfaces in motion (Hay, 1966), pure translatory motion (Clocksin, 1980), planar or fixed-axis motion (Webb and Aggarwal, 1981; Hoffman and Flinchbaugh, 1982; Bobick, 1983), and translation perpendicular to the rotation axis (Longuet-Higgins, 1982). Related studies have shown how certain three-dimensional parameters such as surface Gaussian curvature (Koenderink and van Doorn, 1975), time to collision (Lee, 1976 and 1980) and edge types (Clocksin, 1980) can be extracted from the transforming image. A review of these and other computational results obtained to date on the recovery of structure from motion will appear in (Ullman, 1983).

In summary, the uniqueness results establish that by exploiting a rigidity constraint the recovery of three-dimensional structure is possible on the basis of image motion information alone, and that the recovery is possible in principle on the basis of information that is local in space and time.

7.1.3 Additional Requirements for the Recovery of Structure from Motion

There are a number of interesting differences between the mathematical results cited above and the recovery of structure from motion by the human visual system.

Extension in time: Although the recovery of structure from motion is possible in principle from the instantaneous velocity field, the human visual system requires an extended time period to reach an accurate perception of the three-dimensional structure (Wallach and O'Connell, 1953; White and Mueser, 1960; Green, 1961). This difference is not surprising, especially when the recovery scheme is applied locally to small objects or local surface patches. For surface patches extending about 2° of visual angle, drastically different objects can induce almost identical instantaneous velocity fields. This limitation of the instantaneous velocity field is illustrated in Figure 1. The figure shows a cross-section of two surfaces, $S1$ and $S2$, seen from a side view. The surfaces are assumed to be rotationally symmetric around the observer's line of sight, so that $S1$, for example, is a part of the surface of a sphere. When the viewing distance is such that the surfaces in Figure 1 occupy 2° of visual angle, the difference in their velocity fields within the entire patch does not exceed 6%. The implication is that although the instantaneous velocity field contains sufficient information for the recovery of the three-dimensional shape, the reliable interpretation of local structure from motion requires the integration of information over a more extended time period.

Deviations from rigidity: In interpreting structure from motion, the visual system can cope with less than strict rigidity (Johansson, 1964, 1978; Jansson and Johansson, 1973). If the viewed object undergoes a rigid transformation combined with some non-rigid distortions, the changing shape of the three-dimensional object can often be perceived. This tolerance for deviations from rigidity also implies that the recovery process enjoys a certain immunity to noise. If the viewed object is in fact rigid, but the measurement of its motion and the computations performed are not

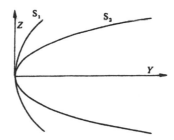

Figure 1. Limitations of the instantaneous velocity field. S_1 and S_2 are two rotationally symmetric surfaces. When the surfaces occupy 2° of visual angle, the difference in their velocity fields at each point does not exceed 6%.

entirely accurate, the result would not be a complete breakdown of the recovery process, but a perception of a slightly distorting object (Petersik, 1979). Robustness with respect to errors in the measured velocity field is particularly important, since an accurate measurement of the velocity field is surprisingly difficult to obtain (Fennema and Thompson, 1979; Horn and Schunck, 1981; Adelson and Movshon, 1982; Ullman and Hildreth, 1983). For human observers, kinetic depth displays of rigid objects often give rise to the perception of a somewhat distorting object (Wallach, Weisz and Adams, 1956; White and Mueser, 1960; Green, 1961; Braunstein, 1962).

Successive approximation: For the human visual system, the recovery of structure from motion is not an all-or-none process. Some notion of a coherent three-dimensional structure can be obtained with short presentations (Lappin, Doner and Kottas, 1980), but an accurate perception of the three-dimensional shape usually requires a longer viewing time (White and Mueser, 1960; Green, 1961). For short viewing times, objects often appear flatter than their correct three-dimensional structure. For example, a rotating cylinder (Ullman, 1979a) would not appear to have the full depth of the correct three-dimensional object. The perceived shape is qualitatively similar to the correct one, and, as mentioned above, often improves with time.

Integrating sources of three-dimensional information: The kinetic depth experiments demonstrated the capacity of the human visual system to recover three-dimensional structure on the basis of motion information alone. Wallach and O'Connell, for example, included in their experiments wireframe objects whose static projection induced no three-dimensional perception. Such objects appear initially as flat and lying in the frontal image plane, but acquire the correct three-dimensional shape when viewed in motion. Under more natural conditions, motion information is seldom dissociated from other sources of information. In this case, problems related to the integration of different sources of information arise. For instance, a kinetic depth experiment has been reported with an object whose static projection was often perceived as a 90° corner, i.e., three mutually perpendicular rods (Wallach, O'Connell and Neisser, 1953). The actual shape of this object was different: the angle between the rods was in fact 110° rather than 90°. Observers who initially saw the 90° corner, often perceived a three-dimensional structure that agreed better with the object's correct structure after viewing the projection of the moving object for sufficiently long time. The initially perceived structure is therefore not necessarily flat, and it may be influenced by different sources of three-dimensional information. In the above experiment the structure from motion interpretation sometimes eventually dominated over the static cues. There are also cases of conflicting three-dimensional information in which the perceived three-dimensional structure is determined primarily by the static cues rather than the motion information (Ullman, 1979, Ch. 5). The case where the structure from motion process is ineffective is not directly relevant to the present discussion, and will not be considered further.

7.1.4 A Hypothesis: Maximizing Rigidity Relative to the Current Internal Model

The above discussion suggests that to be comparable in performance to the human visual system the process of recovering structure from motion should meet the following requirements. (i) At each instant there should exist an estimation of the three-dimensional structure of the viewed object.

This internal model of the viewed structure may be initially crude and inaccurate, and may be influenced by static sources of three-dimensional information. (ii) The recovery process should prefer rigid transformations. (iii) The recovery scheme should tolerate deviations from rigidity? (iv) It should be able to integrate information from an extended viewing period. (v) It should eventually recover the correct three-dimensional structure, or a close approximation to it.

Most of these requirements can be met naturally by the following "incremental rigidity" scheme. Assume that at any given time there is an internal model of the viewed object. Let $M(t)$ denote the internal model at time t. As the object continues to move, its projection would change. If $M(t)$ is not an accurate model of the object at time t, then no rigid transformation of $M(t)$ would be sufficient to account for the observed transformation in the image. The crucial step is that the internal model would then be modified by the minimal change that is still sufficient to account for the observed transformation. In other words, the internal model resists changes as much as possible, and consequently becomes as rigid as possible.

Such a scheme takes into account the use of a current three-dimensional model which initially may be inaccurate (requirement i), the tendency to perceive rigid transformations when possible (requirement ii), without requiring strict rigidity (requirement iii). It also combines information from extended viewing periods, by incorporating incremental changes into the internal model (requirement iv). It thus has the appealing property of combining information from extended periods, and at the same time using at any instant only the internal model and the incoming image at that instant. It does not require storing and using long sequences of different views of the object as might be used, for example, in a computer implementation of a structure from motion process. Unlike Johansson's (1974) trajectory-based scheme, which also integrates information over an extended viewing period, this mode of temporal integration is not limited to fixed-axis motion, but can be applied to objects under general motion.

The incremental rigidity scheme therefore meets four of the five requirements listed above. It remains unclear, however, whether such a scheme can cope with the last, and most crucial, requirement. That is, if $M(t)$ is initially incorrect, would it eventually converge to the correct three-dimensional structure? The answer is not obvious: if the model is incorrect at time t, it is not clear whether an attempt to transform it as rigidly as possible would bring it any closer to the unknown structure of the viewed object. To assess the feasibility of the incremental rigidity scheme it is therefore necessary to examine whether the incremental changes in $M(t)$ would cause it in the long run to converge to the correct three-dimensional structure of the viewed object. The main problem regarding the incremental rigidity scheme can thus be summarized as follows. If $M(t)$ is updated by transforming it at each instant as rigidly as possible, will it converge eventually to the correct three-dimensional structure under rigid motion, and under deviations from rigidity? This problem is examined in the next two sections. Section 7.2 describes more precisely the incremental rigidity scheme and how it is applied to recover three-dimensional structure from motion. Section 7.3 describes the results of applying this scheme to rigid as well as non-rigid objects. The general conclusion is that the incremental rigidity scheme copes successfully with rigid objects as well as with considerable deviations from rigidity, and that it resembles various aspects of the perceptual recovery of structure from motion.

7.2 The Incremental Rigidity Scheme

Analytic treatment of the convergence requirement did not seem tractable; therefore, a computer program was devised to test the convergence of the incremental rigidity scheme to the correct three-dimensional structure under both rigid and non-rigid motion. This section will describe the incremental rigidity scheme that has been employed. Section 7.2.1 describes the basic scheme, 7.2.2 considers possible modifications, and 7.2.3 examines briefly problems of efficiency, including the execution of the scheme in an analog and distributed manner. Section 7.3 then describes the behavior of the scheme as revealed by the computer simulations, and in particular its convergence to the correct three-dimensional structure under rigid and non-rigid motion.

7.2.1 The Basic Scheme

For the computer implementation it is convenient to consider the visual input as a sequence of frames, each one depicting a number of identifiable feature points rather than a continuous flow. The temporal discreteness of the input is not a necessary aspect of the scheme, a continuous formulation is also conceivable. The scheme maintains and updates an internal model, $M(t)$, of the viewed object. $M(t)$ consists of a set of three-dimensional coordinates, (X_i, Y_i, Z_i). Assuming orthographic projection onto the X–Z image plane, (X_i, Z_i) are the image coordinates of the ith point, and Y_i is its depth as estimated in the current model. X–Z was chosen as the image plane, with the positive Y direction pointing away from the observer. This notation keeps the coordinate system right-handed. As will be noted below, a similar scheme can be defined for a perspective rather than orthographic projection. For small objects, or small surface patches of objects, the two projections are in close agreement, consequently the type of projection employed has little effect on the behavior of the scheme. The relation between the two projections is discussed in more detail in Section 7.4.1. In the absence of information about the three-dimensional shape of the viewed object, the initial model $M(t)$ at $t=0$ is taken to be completely flat, i.e., $Y_i=0$ for $i=1...n$, where n is the number of points considered in the computation. The Y_i's could be set to any constant, since the overall distance to the object remains undetermined.

When a new frame corresponding to a later time, t', is considered, the problem is to update $M(t)$ so as to agree with the new frame, while making the transformation from $M(t)$ to $M(t')$ as rigid as possible. The new frame is represented as a set of two-dimensional image coordinates, (x_i, z_i). The new depth values, y_i, are as yet undetermined. It is assumed, however, that the correspondence between points in the two successive frames is known. When the y_i values have been estimated, the set of coordinates (x_i, y_i, z_i) is the estimated structure at time t', denoted by $S(t')$. The notation convention used is that all the parameters that refer to $M(t)$ are denoted by capital letters, e.g., X, Y, Z, etc., and those referring to $S(t)$ by small letters, e.g., x, y, z, etc.

The most rigid transformation of the internal model $M(t)$ is now determined in the following manner. Let L_{ij} denote the distance between points i and j in $M(t)$. That is,

$$L_{ij}^2 = (X_i - X_j)^2 + (Y_i - Y_j)^2 + (Z_i - Z_j)^2. \tag{1}$$

Similarly, l_{ij} is the internal distance in the estimated structure between points i and j at time t'. That is,

$$l_{ij}^2 = (x_i - x_j)^2 + (y_i - y_j)^2 + (z_i - z_j)^2. \tag{2}$$

A rigid transformation implies that $L_{ij}=l_{ij}$, for all i,j, that is, all the internal distances in the object remain unchanged. To make the transformation as rigid as possible, the unknown depth values y_i should therefore be chosen so as to make the values of l_{ij} and L_{ij} agree as closely as possible. If $D(L_{ij}, l_{ij})$ is a measure of the difference between L_{ij} and l_{ij}, then the problem of determining the most rigid transformation of the model can be formulated as determining the values of y_i so as to minimize the overall deviation from rigidity,

$$D(M(t), S(t')) = \sum_{i,j} D(L_{ij}, l_{ij}), \, i = 1, \ldots, n-1, \, j = i+1, \ldots, n.$$

This deviation from rigidity can also serve as a useful "confidence" measure: the smaller the deviation, the more likely is the internal model to
reflect accurately the correct three-dimensional structure of the observed
object.

A reasonable choice of the distance function D should make the contributions from nearby points weigh more than distant ones. The reason is
that the nearest neighbors to a given point are more likely to belong to the
same object than distant neighbors. A point is consequently more likely to
move rigidly with its nearest neighbors. An simple example of a distance
measure with such a falloff is,

$$D(L_{ij}, l_{ij}) = \frac{(L_{ij} - l_{ij})^2}{L_{ij}^3}. \tag{3}$$

As required, the effect in this measure of, say, a 10% change in L_{ij}
decreases as L_{ij} increases.

After the values of y_i have been determined using the minimization
criterion, (x_i, y_i, z_i) becomes the new model $M(t')$. The change in the model
M between time t and t' corresponds to the perceived change in the object
during this time interval. The motion is given in relative terms, and may
arise from either the object's or the observer's motion.

In summary, the computation involved at each step in establishing the
most rigid interpretation is the following. Given an internal model, $M(t)$,
in the form (X_i, Y_i, Z_i), $i=1,...,n$, and the new frame (x_i, z_i), $i=1,...,n$, find a
set of depth values, $\{y_i\}$ such that the overall deviation from rigidity,
$D(M(t), S(t'))$ is minimized.

7.2.2 Possible Modifications

Some modifications of the basic scheme presented above are possible. For example, a somewhat different form of the metric **D** can be used. The important issue to explore, however, is whether any such scheme converges successfully to the correct three-dimensional structure. As discussed above, the incremental rigidity scheme meets requirements (i) through (iv), but it is unclear whether it can also meet the convergence requirement. To be considered a plausible scheme for the recovery of structure from motion by the human visual system, the convergence requirement must also be met for rigid as well as not strictly rigid objects. If a particular version of the scheme accomplishes the three-dimensional recovery task successfully, then it provides a certain existence proof that an incremental rigidity scheme can meet all of the requirements listed in Section 7.1.4.

Two modifications of the basic scheme described above were explored. One was to introduce some changes to the metric **D**. The other, more substantial modification, takes into account the fact that $M(t)$, the internal model at time t, may be inaccurate, and allows it to be corrected.

The basic scheme described in the previous section can be summarized as minimizing $\mathbf{D}(M(t),S(t'))$, a measure of the overall distortion between the three-dimensional model at time t and the new computed structure at time t'. The modified method searches for a modified, corrected model $M'(t)$, such that the transition from $M(t)$ to $M'(t)$ (the correction to the internal model) is small, and the transition from $M'(t)$ to $S(t')$ is as rigid as possible. This modified scheme minimizes therefore the sum

$$\mathbf{D}(M(t), M'(t)) + \mathbf{D}(M'(t), S(t')).$$

Since it allows changes in the internal model $M(t)$, this scheme will be referred to below as the "flexible model" scheme. In general, the modifications explored of the metric **D** had only small effects on the convergence to the correct three-dimensional structure. The use of the more complicated flexible model scheme also did not introduce fundamental changes, but usually resulted in an overall improvement of the computed structure. This flexible model also has the advantage that other three-dimensional cues, such as stereo or shading, could influence the transition from $M(t)$ to $M'(t)$ The above observations suggest that the basic incremental rigidity scheme is not sensitive to variations in the exact formulation of the minimization problem. Additional comments regarding the modified scheme are incorporated in the discussion of the results in Section 7.3.

7.2.3 Implementation

The incremental rigidity scheme described above has been implemented as a computer program on a Lisp Machine at the MIT Artificial Intelligence Laboratory. The computation made use of a relatively efficient variable-metric minimization procedure (Davidon, 1968). For a quadratic function of n variables, this method is guaranteed to converge to a minimum within at most n iterations. The computational load at each stage is relatively small, estimated to require approximately $\dfrac{3n^2}{2}$ multiplications (Davidon, 1968). When the objective function, in our case, D, has more than a single minimum, the minimization process will converge to a local, but not necessarily the global, minimum. The results described in Section 7.3 demonstrate that this convergence is sufficient for the recovery of the unknown three-dimensional structure. Some consequences of the convergence to the local minimum are discussed in Section 7.4.4.

For the flat initial model, an additional minor step is required to ensure convergence to a local minimum. The flat internal model can change into two equally likely configurations, one being the mirror image reflection of the other about the image plane. The model is therefore perturbed slightly, to cause it to prefer one of the two symmetric minima over the other.

This minimization method is efficient for implementation on a serial digital computer. More parallel distributed implementations are also possible. Such extensions will not be analyzed here mathematically for a discussion of minimization in distributed networks see (Ullman, 1979b). Instead, a mechanical analogue that performs essentially the same computation in a parallel distributed manner will be briefly described. This mechanical spring model, which bears some similarity to Julesz' spring-dipole model of stereopsis (Julesz, 1971), can help to visualize the computation performed at each step by the incremental rigidity scheme, and can be helpful in suggesting parallel distributed computations for maximizing rigidity.

The mechanical spring model is illustrated in Figure 2 for a three element object. As before, let (x_i, z_i) denote the image coordinate of the ith point, and y_i be the unknown depth coordinate that must be recovered. This situation is modeled in Figure 2 by a set of rigid rods, each one connected to one of the viewed points and extending perpendicular to the image plane. The ith point is constrained to lie along the ith rod, but its position along this rod, i.e., its depth value, is still undetermined. The points are now connected by a set of springs. The resting length of the spring connecting points i and j is L_{ij}, their distance in the internal model prior to the introduction of the new frame, and k_{ij} is the spring constant. The points will now slide along the rods, stretching some of the springs and compressing others, until a minimum energy configuration is reached. If l_{ij} denotes the distance between points i and j in the final configuration, then the total energy of the system would be

$$E(M(t), S(t')) = \frac{1}{2} \sum_{i,j} K_{ij}(L_{ij} - l_{ij})^2.$$

To mimic the computation described in the preceding section, the spring constants, k_{ij}, should be smaller for longer springs. It can also be assumed that each point is connected only to a number of its nearest neighbors. The "computation" of the most rigid transformation is performed in this mechanical system in a parallel distributed manner. It can be used, therefore, to illustrate the possibility of maximizing rigidity in the observed transformation using a parallel network of simple interacting computing

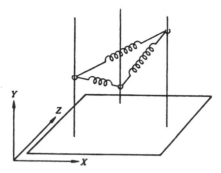

Figure 2. A spring model for the distributed computation of the most rigid interpretation. Each of the viewed points (three in this example) is constrained to move along one of the rigid rods, and its position along the rod represents its depth value. The connecting springs represent the distances between points in the current internal model. The points would slide along the rods until a minimum energy configuration is reached. The final configuration represents the modified internal model.

elements.

This mechanical analogue illustrates the computation for the case of orthographic projection. For perspective projection, only a slight modification is required: the rods should converge to a common point rather than be perpendicular to the image plane. Continuous versions of this scheme, in which the rods move continuously and the springs' lengths and constants are also modified continuously are possible, but they will not be discussed further here.

7.3 Experimental Results

7.3.1 Rigid Motion

Typical results showing the incremental rigidity scheme in operation are illustrated in Figure 3. The object in this example is shown in Figure 3a It contains six points: the vertices of the outlined pentagon, and a sixth point at the origin, marked by the unfilled circle. The object was three-dimensional, not merely planar. The two-dimensional projection of this object is shown in the figure from a top view, i.e. as projected on the X–Y plane. The input to the incremental rigidity program consisted of the projection of the object on the X–Z image plane. That is, only the (x_i, z_i) coordinates for $i=0, \cdots, 5$ were given. This projection on the X–Z image plane at time $t=0$ is shown in Figure 4. Clearly, such an image conveys no information regarding the object's correct three-dimensional structure. The unknown depth values were assumed initially to be constant, $y_i=0$ for $i=0, \cdots, 5$. That is, no depth was assumed, and the initial internal model consisted of a planar object, lying parallel to the image plane. The dashed line in Figure 3a illustrates the projection of the internal model onto the X–Y plane.

The object was then rotated by 10° at a time, and the internal model was modified according to the scheme described in Section 7.2.1. The rotations were around the vertical Z axis. Any other axis in space can be used instead, however, and the axis may also change over time. To illustrate the behavior of the scheme, the error between the internal model and the object's correct three-dimensional structure was computed at the end of each step. This error was measured by the root-mean-square expression,

$$E(O, M(t)) = \frac{1}{2} \sum_{i,j} K_{ij}(d_{ij} - L_{ij})^2.$$

where d_{ij} is the correct three-dimensional distance between points i and j in the object, O, and L_{ij} is the corresponding distance in the internal model,

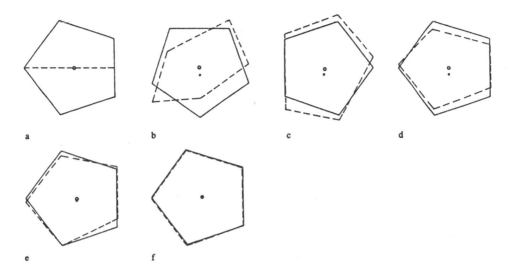

Figure 3. The incremental rigidity scheme applied to a six-point rigid object (the solid outlined pentagon and unfilled dot at the center). The object is three-dimensional, the figure depicts its projections on the X-Y plane (Y is the depth axis). The internal model (dashed curves and filled dot near the center) is compared to the correct structure following (a) 0 rotation, (b) 90°, (c) 180°, (d) 360°, (e) 2 rotations, (f) 4 rotations.

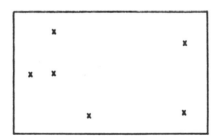

Figure 4. The initial projection of the 6-point object on the image (X-Z) plane.

$M(t)$. When the internal model is completely accurate, $L_{ij}=d_{ij}$ for all i,j and the total error vanishes. The initial error was normalized to 1.0 at 0° rotation. Its development as a function of the rotation angle is shown in Figure 5 (dotted curve). It declines rapidly to about 0.3 after the first 180° of rotation, (note in Figure 3c that this error already yields an approximation to the actual structure) and then continues to decline to about 2–5%. Figure 3 shows the development of the internal model. It starts as entirely flat (3a). After 90° of rotation it acquires some depth, but it is still too flat, and the shape is inaccurate (3b). After the first 180° of rotation the internal model is already similar in overall shape to the correct structure (3c). The internal model continues to improve (3d and e) until it becomes virtually indistinguishable from the correct three-dimensional structure (3f).

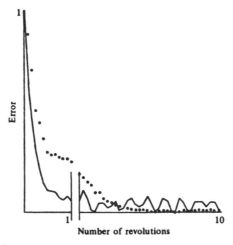

Figure 5. The error of the internal model as a function of rotation angle over 10 revolutions. The initial error for the basic scheme (dots) and flexible model (solid curve) is normalized to 1. The error is shown every 30° for the first revolution, and every 90 ° afterwards.

A more rapid initial approximation to the correct three-dimensional structure can be obtained by using the flexible model scheme described in Section 7.2.2. The continuous curve in Figure 5 illustrates the error measure as a function of rotation angle for the flexible model scheme. It can be seen that the approximation improves rapidly over the first 180° of rotation, but it remains somewhat more oscillatory than the basic scheme.

The results of applying the incremental rigidity scheme to various objects in motion show that for most of the rotation time the internal model approximates the actual three-dimensional structure. The model does not converge, however, to the precise solution, but often wobbles somewhat around the correct solution to the three-dimensional structure. In both the basic and the flexible model schemes the approximation to the correct solution does not improve monotonically as a function of rotation angle. The lack of monotonicity in the overall convergence to the computed three-dimensional structure suggests that an analytic treatment of the convergence properties of the incremental rigidity scheme is probably difficult to obtain.

During these oscillations of the error function, the correct structure is occasionally lost temporarily and then recovered. In the course of such a phase, when the structure is lost and recaptured, a spontaneous depth reversal may occur. That is, the internal model converges not to the original three-dimensional structure, but to its mirror image, reflected about the image plane. The convergence to the reflected rather than to the correct structure is a "legal" solution under orthographic projection, since, as noted in Section 7.1.2, the two are indistinguishable under orthographic projection. This point of depth reversals under orthographic and perspective projections is discussed further in Section 7.4.1.

Such a depth reversal is shown in Figure 6, which illustrates the development of the internal model for a five point object, similar to the six point object examined before, but without the point at the origin. Figures 6a, b, and c. compare the internal model, shown as dashed lines, with the actual structure, the solid lines, following 0°, 90°, and 180°, respectively. Towards the end of the second revolution the structure was temporarily lost, and then recovered successfully. During this phase, a depth reversal occurred. That is, the internal model later converged not to the correct three-dimensional structure, but to its mirror image, reflected about the image plane. The approximation to the reflected structure eventually becomes quite accurate. Figure 6d shows the correct structure together with the best approximation obtained within the first five revolutions. Figure 6e is similar to 6d, but the correct structure has been reflected about the image plane. It can be seen that the internal model provides a good approximation to the reflected structure. The best approximation obtained within the first ten revolutions is compared in Figure 6f against the correct, but reflected, structure. That the structure is recaptured following a total loss, together with the initial convergence from a totally flat internal model, indicates that almost irrespective of the initial conditions the scheme eventually converges, in the sense that it spends most of its time near the correct solution.

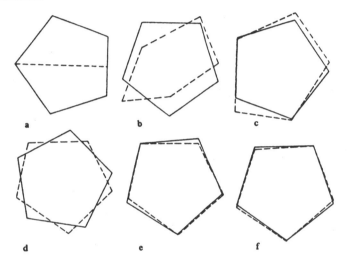

Figure 6. The internal model (dashed figures) is compared to the correct structure (solid pentagons) following (a) 0°, (b) 90°, (c) 180°, (d-e) 5 revolutions, (f) 10 revolutions. Towards the end of the second revolution, the structure was temporarily lost. During this phase, a depth reversal occurred (d). Figures (e) and (f) compare the internal model to the correct three-dimensional structure reflected about the image plane.

In the examples above the objects have been rotated 10° between successive views. It might have been expected that if a sequence of frames is taken, say, every 5° of rotation instead of the 10° used above, the recovery of the structure would require a smaller overall rotation, since the deviation from rigidity at each step is smaller. In fact, when smaller angular separations between views are used, the convergence becomes somewhat slower. Figure 7 compares the decline of the error function over the first five rotations for 10°, shown as a dotted curve, and 5°, the continuous curve, rotations between successive views. This difference in convergence rate suggests that the incremental rigidity scheme performs better when successive views of the object differ significantly. This preference may be related to the findings of Petersik (1980) who compared the contribution of the short- and long-range motion processes (Braddick, 1974) to the recovery of structure from motion. The long-range process, which operates over relatively large spatial and temporal separations, was found in this study to be the main contributor to the structure from motion process.

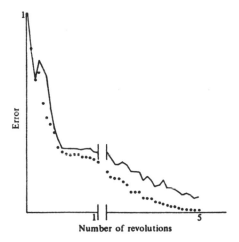

Figure 7. The decline of the error function for a rotation rigid object when successive frames are separated by 5° (solid) and 10° (dotted curve) of rotation. The convergence is faster when successive views of the object are sufficiently distinct.

Comparisons were also made with the type of to-and-fro motion used in the original kinetic depth experiments by Wallach and O'Connell. In the examples examined above, the objects rotated continuously in one direction for several rotations. In contrast, the objects in Wallach and O'Connell's experiments were rotated to-and-fro through a limited angular excursion.

Under this condition, the observers did not have the benefit of viewing the object from all directions, but they were nevertheless able to recover the correct three-dimensional structure of the moving objects. A simulation of this condition, in which the objects were rotated by only 40° in each direction, revealed that the incremental rigidity scheme manifests a similar capacity. As the object rotated to-and-fro, the internal model continued to improve until the correct three-dimensional structure was recovered, in a manner analogous to the recovery of the three-dimensional structure of continuously rotating objects.

In summary, when applied to rigid objects in motion, i.e., the objects described in Section 7.3, 7.4 and others, the incremental rigidity scheme exhibits the following properties. (1) Veridicality: for most of the time a reasonable approximation of the correct three-dimensional structure is maintained. (2) Temporal extension: the time, i.e., number of frames, required for an approximation to be obtained is longer than the theoretical minimum required for the recovery of structure from motion (Ullman, 1979, Longuet-Higgins and Prazdny, 1980; Tsai and Huang, 1982). (3) Residual non-rigidity: although the changing image is induced in this case by completely rigid objects in motion, the computed three-dimensional structure included residual non-rigidity. (4) Non-monotonicity: Starting from a flat internal model, the solution generally improves with time, however, phases of increasing error recur in the extended analysis. (5) Depth reversals: occasionally the increased error is associated with a spontaneous depth reversal. The flexible model scheme is less susceptible to such reversals.

Similar general properties are also manifested in the perception of structure from motion by human observers. The perceived three-dimensional structure is usually similar to the correct three-dimensional structure. It improves with time, but it is usually not entirely accurate (Wallach and O'Connell, 1953; White and Mueser, 1960). The perception is often of a stable three-dimensional configuration accompanied by some residual elastic deformations, particularly when the number of participating elements is small.

7.3.2 Non-Rigid Motion

In this section the capacity of the incremental rigidity scheme to cope with deviations from rigidity will be examined. Unlike the previous section where the viewed objects were assumed to be entirely rigid, in this section they are allowed to deform while they move.

An example of the scheme applied to non-rigid motion is shown in Figure 8. At time $t=0$ the object was identical in shape to the five point object examined under rigid motion in the previous section. Again, the object was three-dimensional and not merely planar. The figure depicts the projection of this object from a top view on the X-Y plane. A non-rigid transformation was added to the rotation of the object. The non-rigid distortion was quite significant, as can be seen in Figure 8a. The shape of the object following two revolutions is compared in the figure with its original shape. The incremental rigidity scheme copes successfully with such deviation from rigidity. The internal model by the end of the second revolution is shown in Figure 8b for comparison with the correct structure.

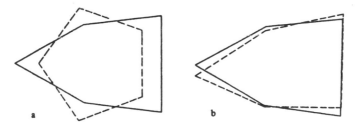

Figure 8. The recovery of non-rigid shape. A pentagon distorts while it moves. (a) Its shape following 2 revolutions (solid) is compared with its original shape (dashed lines). (b) The internal model by the end of the second revolution (dashed) is compared with the correct three-dimensional structure seen from a top view (solid lines).

Figure 9 illustrates the results of applying the same scheme to an

object distorting twice as fast. That is, the distortion of the object after one
revolution is identical to the distortion spread in the previous example over
two revolutions. The figure shows the actual object with its internal model
after 180° (9a), 360° (9b), 450° (9c), and 720° of rotation (9d). The error
at 450° is relatively low, but at 720° the error is relatively high again.
Note that while the internal model for the example illustrated in Figure 9
becomes less accurate compared to the lower distortion rate example of
Figure 8, the three-dimensional structure is still essentially recovered.

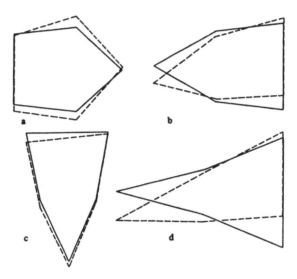

Figure 9. An object distorting twice as fast as that in Figure 8. The
internal model (dashed line) is compared to the correct structure following (a) 180°, (b) 360°, (c) 450°, (d) 720° of rotation. When the rate of
distortion increases further, its structure can no longer be recovered by
the incremental rigidity scheme.

When the rate of distortion was doubled again, the incremental rigidity scheme could no longer cope with the rate of deviation from rigidity.
Before the end of the second revolution the structure was lost entirely. The
distortion was evidently developing too fast to allow the scheme to recover
from this loss. This limitation held for both the basic and the flexible
model schemes. In contrast with previous cases, the error measure in this

case tends to grow without bounds.

Different objects under different distortions were also examined, with similar results. For moderate distortions the incremental rigidity scheme can cope successfully with non-rigid motion. The amount of distortion that can be tolerated is difficult to quantify, but as illustrated in Figures 8 and 9 it can be substantial.

As noted in Section 7.1.3, the human visual system can also cope to some degree with kinetic depth effects that are not entirely rigid. Although no systematic studies of this capacity have been reported, the human visual system is probably susceptible to similar difficulties with non-rigid motion, and it should be expected to fail under pure kinetic depth conditions, i.e., when no other sources of three-dimensional information are available, when the deviation from rigidity becomes excessive.

7.4 Additional Properties of the Incremental Rigidity Scheme

This section discusses four additional topics pertaining to the incremental rigidity scheme under both rigid and non-rigid motion. Within each topic, the scheme is compared to previous mathematical models and to human perception of structure from motion.

7.4.1 Orthographic and Perspective Projections

The computations described above used orthographic projection. As noted in Section 7.2.3, a similar scheme can also be adapted for perspective projection. For objects that are small in comparison to the overall viewing distance the two projections are similar. It is not surprising, therefore, that although perspective effects can be useful under some conditions (Braunstein, 1962, 1976; Lappin and Fuqua, 1983), the human visual system can recover structure from motion under both perspective and orthographic projections. For the kinetic depth effect, it has been reported that perspectivity has a very limited effect on the perception of objects

subtending up to about 15° of visual angle (Johansson, 1978). Under such
conditions orthographic projection can be viewed as slightly distorted perspective projection, and since the recovery scheme should be insensitive to
small distortions, though it should be able to cope with both types of projection. In fact, the capacity to deal successfully with both types of projection can be used as a test for the scheme's robustness. A scheme that can
recover the structure under perspective projection but fails under orthographic projection cannot be robust when applied locally. This comment is
relevant in particular to schemes that rely exclusively on the instantaneous
velocity field, since such schemes fail under orthographic projection (Ullman, 1983).

For larger objects, though perspective and orthographic projections
differ substantially, it is still possible, to use a parallel scheme (Ullman,
1979) in which the interpretation is performed locally. Therefore it is
immaterial whether orthographic of perspective projections are employed,
and the local results can be combined then in a second stage. For
sufficiently large objects this integration stage would eliminate the ambiguity inherent in orthographic projection regarding direction of rotation and
reflection about the image plane. In conclusion, for small objects or surface
patches it is immaterial which of the two projection types is employed, and
a robust recovery method should be able to cope with both. For larger
objects it is still theoretically possible to use either type, and either one can
be incorporated in the incremental rigidity scheme.

7.4.2 The Effect of the Number of Points

In this section the effect of the number of moving feature points will
be discussed by comparing its application to two, three, four, and many
points in motion.

Two points in motion: Mathematically, for two points the three-
dimensional structure is not determined uniquely by any number of views.
The structure imposed by the incremental rigidity scheme would be of a

rigid rod rotating in depth, and the view where the rod's length is maximal would be taken as lying in the frontal plane. This three-dimensional interpretation is in agreement with human perception of two-dot configurations (Johansson and Jansson, 1968).

Three points: This configuration has not been analyzed mathematically in the past. It is known that four points in three views determine the three-dimensional structure uniquely if the structure is assumed to be rigid. Three points in three views do not always guarantee uniqueness, but it is still possible that with additional views the three-dimensional structure is determined uniquely by three points alone. The results of applying the incremental rigidity scheme support this possibility, since the three-dimensional structure of three-point configurations can be successfully recovered. This result holds for orthographic projections. For perspective projections, three points are still insufficient.

The recovery of the three-dimensional structure of three moving points is shown in Figure 10. As before, the initial model was taken as entirely flat. The evolving internal model, shown by the dashed line, can be compared in the figure with the actual three-dimensional structure following 90°, 180°, 360° and 720° of rotation. The figure shows that a fast and accurate recovery can be obtained for only three points in motion.

Four points: Three views are theoretically sufficient for recovering the structure of four non-coplanar points moving rigidly. The incremental rigidity scheme can recover the structure of four point objects, but three views are insufficient for an accurate recovery.

The recovery of the three-dimensional structure of four points arranged in a square when viewed from above is shown in Figure 11. The initial model was taken again as entirely flat. Again the internal model is shown by the dashed line following 90°, 180°, 360° and 720° of rotation and can be compared to the actual three-dimensional structure. It can be seen that the model is initially inaccurate, but that the correct three-dimensional structure is eventually recovered and retained with only minor

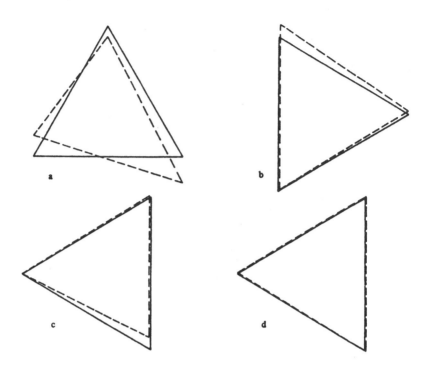

Figure 10. The incremental rigidity scheme applied to a 3-point object. The internal model (dashed lines) is compared to the correct structure (solid lines) as seen from a top view, following (a) 90°, (b) 180°, (c) 360° and (d) 720° of motion. Three points are sufficient for the recovery of structure from motion.

residual deviations from rigidity.

Additional points: The effect of additional points on the incremental rigidity scheme would depend on the method of applying the scheme to large collections of elements. There are two possible methods of applying the scheme to such large collections. The first is a single-stage scheme, in which the computation described in Section 7.2 is simply applied simultaneously to all of the elements in view. The second possibility is a two-stage scheme similar to my polar-parallel scheme (Ullman, 1979a). In the first stage the incremental rigidity scheme is applied independently to small subcollections of elements. In the second stage the local results are

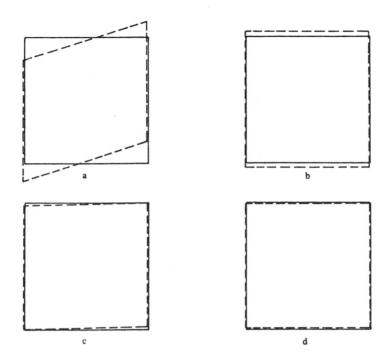

Figure 11. The recovery of the three-dimensional structure of 4 points arranged in a square when seen from a top view. The initial model was entirely flat. The model (dashed) is compared to the actual structure following (a) 90°, (h) 180 °, (c) 360°, and (d) 720° of rotation. The recovery takes longer than the known minimum of 3 views, but the correct structure is eventually recovered and maintained.

combined. It is expected that for the two-stage scheme there would be a more noticeable improvement with the number of elements (c.f. Braunstein, 1962; Johansson, 1978).

The effects of numerosity would also depend on the function, **D**, used in measuring the deviation from rigidity (Section 7.2). If it falls off rapidly as a function of spatial distance, only the nearest neighbors would make substantial contribution to the computation, and the effect of numerosity would be more restricted compared to a function that falls off more gradually with distance.

7.4.3 On Multiple Objects

One possible method of dealing with multiple independently moving objects is to segregate the scene into objects, e.g. on the basis of distance, common two-dimensional motion characteristics, etc., before applying the rigidity-based interpretation scheme to each object separately. An interesting alternative is that object segregation may not be required as a separate stage, but may be a by-product of the three-dimensional interpretation process. Similar to the previous section, two methods for achieving such segregation seem possible. First, the segregation may be accomplished in a single stage process by the appropriate choice of the deviation measure, **D**. The suggestion is that **D** would prefer "partial rigidity" in the following sense. Suppose that no rigid transformation of the internal model can account for the incoming input. The model must then be modified non-rigidly. For simplicity, assume that only two different modifications of the model are possible. In the first, all the internal distances are changed somewhat. The second maintains partial rigidity: some distances change more than in the first type of modification, but others remain completely rigid. We would want the measure of deviation from rigidity to be lower for this second, partially rigid, transformation. For two independently moving objects, a scheme that maximizes rigidity would then prefer the solution in which the scene contains two rigid substructures. In this manner, the appropriate choice of **D** may endow the incremental rigidity scheme with the capacity to divide the scene into its rigid components.

A second possibility for dealing with multiple objects is within the framework of the two-stage process mentioned above. The general suggestion is that substructures that share similar motion parameters, e.g., same axis of rotation, in the first stage, will be grouped together in the second stage placing the burden of the segregation process on this latter integration stage.

7.4.4 Convergence to the Local Minimum

The schemes described in Sections 7.2.2 and 7.2.3 seek the local minimum in the measure of deviation from rigidity. As illustrated by the examples examined in Section 7.3, this convergence to the local minimum is usually sufficient to recover the correct three-dimensional structure. Under certain conditions, however, the incremental rigidity scheme may converge to a local minimum which is not the most rigid structure possible. Similar behavior is also exhibited by the human perception of structure from motion. Under certain conditions human observers perceive non-rigid structure in motion when an entirely rigid solution is also possible. A well known example of this phenomenon is the Mach illusion. Mach originally described a static version of this illusion, while the dynamic version was described in (Eden, 1962; Lindsay and Norman, 1972). This illusion can be created by folding a sheet of paper to create a vertical V-shaped figure. Under monocular viewing, this shape is ambiguous and can reverse in depth. To observe the dynamic illusion the observer waits for a depth reversal to occur, and then moves their head in different directions. Under these conditions the object is seen to move whenever the observer's head moves. The illusory motion arises despite the observer's knowledge of the correct three-dimensional configuration, and it often contradicts shading clues, stability criteria, and touch cues (Eden, 1962). When the object is close to the observer's eye, its motion is no longer rigid, but appears to distort considerably while it moves.

The incremental rigidity scheme would also be susceptible to this illusion. The reason lies in the initial internal model. Unlike the pure kinetic depth situation, a three-dimensional structure is perceived from the static view. Because of the depth reversal, the initial internal model resembles the reflection of the three-dimensional structure about the image plane. It subsequently converges not to the correct three-dimensional structure, but to its mirror image which, under perspective projection, can be considerably less rigid than the correct structure.

It has been shown that when one face of a rotating wire cube is increased in brightness, the cube is often perceived as non-rigid (Sperling, et al., 1983). This behavior is again consistent with the incremental rigidity scheme. The brighter face is usually perceived as somewhat closer to the observer even when in fact it may be the fartherest. This bias of the internal model will cause the incremental rigidity scheme to converge to the reflected structure and miss the correct, entirely rigid, solution.

7.5 Possible Implications to the Long-Range Motion Correspondence Process

For the human visual system to perceive motion, the moving object is not required to move continuously across the visual field. Under the appropriate spatial and temporal presentation parameters, a stimulus presented visual system can "fill-in" the gap in the discrete sequence of presentations even when the stimuli are separated by several degrees of visual angle and by long temporal intervals, e.g., 400 ms and more (Neuhaus, 1930).

This phenomena, termed "apparent" or "beta" motion, raises the question of whether discrete and continuous motion are registered by a single mechanism within the visual system. Psychophysical evidence accumulated in recent years supports the notion that two different processes are involved in the detection and measurement of visual motion (Braddick, 1974, 1980).

The two processes have been termed the "short-range" and "long-range" motion detection processes. The long-range process is also called the correspondence process in motion, since it appears to operate by first identifying features, such as small blobs, endpoints, junctions, in the image, and then matching these features over time. A question that arises from these observations is what could be the role of the long-range correspondence process in normal motion perception. The motion of objects in the visual field is almost always continuous, discrete "jumps" over large spatial and temporal gaps are unlikely. One possible line of explanation that has been offered in the past is that the long-range process is needed to deal with occlusions. Moving objects can disappear and reappear behind an occluding object, such as an animal running behind a tree. This explanation appears unlikely, since in this case the long-range motion process in fact does not operate in its normal mode. Unlike the phenomenon of apparent motion, the visual system does not fill-in the trajectory of the occluded object.

A different explanation has been suggested according to which the short-range process and the long-range process are subserving distinct functional roles (Marr, 1982, Ch. 3.5). The short-range motion is used as input for the structure from motion computation, and the long-range process is used to maintain the object identity of non-rigid objects. The results described in this paper suggest the opposite point of view. The results, e.g. Figure 7, suggest that in order to obtain a reliable recovery of three-dimensional structure from motion it is important to compare views of the same object that differ significantly. A comparison of two views separated by 10° of rotation, for example, yields better results than using two comparisons, each restricted to 5° rotation. Assuming that this is a general phenomenon we may conclude that it is essential for the structure from motion computation to be able to compare views of the object that are separated by significant spatial and temporal intervals.

The view I suggest is, therefore, the following: The main input to the structure from motion computation is the long-range correspondence process that can track the precise displacement of well localized features over large spatial and temporal intervals. Since long-range correspondence is often ambiguous and difficult to establish (Ullman, 1979a), it is likely that the short-range process is used to guide and constrain the long-range matches. The proposed view is, therefore, not of two processes working independently and subserving different roles, but of two processes complementing each other and interacting to produce the final measurement of visual motion. The long-range matches obtained in this manner then serve as the main input to the structure from motion computation.

Two observations can be raised in support of this general view, one psychophysical, the other computational. The psychophysical evidence that supports the notion that the long-range process is more critical to the perception of structure from motion (Petersik, 1980).

The computational argument relies on two facts. First, that the recovery of structure based on the instantaneous velocity field requires very precise measurements (Figure 1). The second is that obtaining velocity measurements with such precision appears difficult to obtain (Hildreth, 1984).

7.6 Summary

This chapter suggests a new scheme for the recovery of three-dimensional structure from rigid and non-rigid motion. According to this scheme an internal model representing the three-dimensional structure of the viewed object is maintained and modified as the object moves with respect to the viewer, or changes its structure. The transformations in the internal model thus mirror the changes in the environment, similar to the manner suggested by Shepard (Shepard and Metzler, 1971; Shepard and Cooper, 1982). The internal model resists changes in its shape as much as possible. Consequently, of all the modifications of the model that may account for the observed transformation, the most rigid one is preferred.

This method of recovering three-dimensional structure from motion shares a number of properties with the perception of kinetic depth displays by human observers. The internal model is initially inaccurate and improves with time (successive approximation). In the absence of static three-dimensional information, the model prior to the beginning of the motion may initially be entirely flat. As the object starts to move, the model begins to acquire depth, and eventually it reaches a configuration that approximates the actual three-dimensional structure (convergence). If the initial view does convey three-dimensional information, this information may affect the structure of the internal model (integrating sources of information). The recovery process eventually integrates information from different views of the object (temporal extension). The entire history of the process is summarized, however, in the structure of the internal model: the scheme does not operate on long sequences of views or stored trajectories.

The proposed incremental rigidity approach has raised two main questions: (1) Does the process convergence to the correct three-dimensional structure? (2) Does the scheme have the capacity to cope with deviations from rigidity? The computer simulations have demonstrated that the use of the instantaneous internal model alone, coupled with a principle of maximizing rigidity, is sufficient for the recovery of three-dimensional structure from motion. The resulting scheme has an inherent preference for rigid transformations, but it can also cope with considerable deviations from rigidity.

The simulations also revealed the main advantages and disadvantages of the incremental rigidity scheme. Compared with previous approaches, one limitation of the scheme is that the resulting three-dimensional structure is usually not entirely accurate, although it often approximates the correct structure quite closely. Even for strictly rigid objects, the computed three-dimensional solution usually contains residual non-rigid distortions. On the other hand, two advantages of the incremental rigidity scheme make it an attractive approach to the recovery of structure from motion. The first is its capacity to cope with non-rigid motion: the three-dimensional structure can be approximated in the face of substantial deviations from rigidity. The second is its robustness: errors in the measured velocity and in the computations employed do not result in complete failure, but in some additional non-rigid distortions superimposed on the correct three-dimensional structure.

A number of problems remain open for further studies. Mathematically, it would be of interest to analyze the convergence properties of the incremental rigidity scheme. As noted in Section 7.3.1, one complicating factor in this analysis is the lack of monotonicity in the decline of the residual error. This error function has been defined above as a root-mean-square measure, using the differences in the internal distances in the correct three-dimensional structure and the internal model. It remains possible that a different error measure would prove more amenable to analytic treatment.

From a psychological point of view, there is a qualitative similarity between the perception of structure from motion by humans, and the behavior of the incremental rigidity scheme. Quantitative data regarding the perception of structure from motion under deviations from rigidity are, however, scant. It would be of interest to investigate further this capacity of the human visual system, and compare the empirical results with the behavior of the incremental rigidity scheme.

Acknowledgements

I thank Dr. Ellen Hildreth for her valuable comments and Carol Jean Bonomo for her help in preparing this manuscript. This report describes research done at the Artificial Intelligence Laboratory of the Massachusetts Institute of Technology. Support was provided by National Science Foundation Grant 79-23110MCS. Material in this chapter is reproduced with permission from *Perception*, vol. 13, pp. 255-274.

References

Adelson, E.H., and Movshon, J.A., (1982) 'Phenomenal coherence of moving visual patterns,' *Nature,* vol. 300, pp. 523-525.

Bobick, A., (1983) 'A hybrid approach to structure-from-motion,' *ACM Interdisciplinary Workshop on Motion: Representation and Perception,* Toronto, Canada, pp. 91-109.

Braddick, O.J., (1974) 'A short-range process in apparent motion,' *Vision Research,* vol. 14, pp. 519-527.

Braunstein, M.L., (1962) 'Depth perception in rotation dot patterns: Effects of numerosity and perspective,' *J. of Experimental Psychology,* vol. 64, no. 4, pp. 415-420.

Braunstein, M.L., (1976) **Depth Perception Through Motion,** Academic Press, New York.

Clocksin, W.F., (1980) 'Perception of surface slant and edge labels from optical flow: A computational approach,' *Perception,* vol. 9, no. 3, pp. 253-269.

Davidon, W.C., (1968) 'Variance algorithm for minimization,' *The Computer J.,* vol. 10, no. 4, pp. 406-413.

Eden, M., (1962) 'A three-dimensional optical illusion,' *Quarterly Progress Rep.,* no. 64, MIT RLE, pp. 267-274.

Fennema, C.L., and Thompson, W.B., (1979) 'Velocity determination in scenes containing several moving objects,' *Computer Graphics and Image Processing,* vol. 9, pp. 301-315.

Gibson, J.J., and Gibson, E.J., (1957) 'Continuous perspective transformations and the perception of rigid motion,' *J. of Experimental Psychology,* vol. 54, no. 2, pp. 129-138.

Green, B.F., (1961) 'Figure coherence in the kinetic depth effect,' *J. of Experimental Psychology,* vol. 62, no. 3, pp. 272-282.

Hay, C. J., (1966) 'Optical motions and space perception - An extension of Gibson's analysis,' *Psychological Review,* vol. 73, pp. 550-565.

Hoffman, D.D., and Flinchbaugh, B.E., (1982) 'The interpretation of biological motion,' *Biological Cybernnetics,* vol. 42, pp. 195-204.

Horn, B.K.P., and Schunck, B.G., (1981) 'Determining optical flow,' *Artificial Intelligence,* vol. 17, pp. 185-203.

Jansson, G., and Johansson, G., (1973) 'Visual perception of bending motion,' *Perception,* vol. 2, pp. 321-326.

Johansson, G., (1964) 'Perception of motion and changing form,' *Scandinavian J. of Psychology,* vol. 5, pp. 181-208.

Johansson, G., (1974) 'Visual perception of rotary motion as transformation of conic sections - A contribution to the theory of visual space perception,' *Psychologia,* vol. 17, pp. 226-237.

Johansson, G., (1975) 'Visual motion perception,' *Scientific American,* vol. 232, no. 6, pp. 76-88.

Johansson, G., (1978) 'Visual event perception,' in R. Held, H.W. Leibowitz, and H.-L. Teuber (eds.), **Handbook of Sensory Physiology,** Springer-Verlag, Berlin.

Johansson, G., and Jansson, G., (1968) 'Perceived rotary motion from changes in a straight line,' *Perception and Psychophysics,* vol. 4, no.

3, pp. 165-170.

Julesz, B., (1971) **Foundation of Cyclopean Perception,** University of Chicago Press.

Koenderink, J.J., and van Doorn, A.J., (1975) 'Invariant properties of the motion parallax field due to the motion of rigid bodies relative to the observer,' *Optica Acta,* vol. 22, no. 9, pp. 773-79.

Lappin, J.S., Doner, J.F., and Kottas, B.L., (1980) 'Minimal conditions for the visual detection of structure and motion in three dimensions,' *Science,* vol. 209, pp. 717-719.

Lappin, J.S., and Fuqua, M.A., (1983) 'Accurate visual measurement of three-dimensional moving patterns,' *Science,* vol. 221, pp. 480-482.

Lee, D.N., (1976) 'A theory of visual control of braking based on information about time to collision,' *Perception,* vol. 5, pp. 437-459.

Lee, D.N., (1980) 'The optic flow field: The foundation of vision,' *Proc. Royal Society London,* vol. B 290, pp. 169-179.

Lindsay, P.H., and Norman, D.A., (1972) **Human Information Processing,** Academic Press, New York.

Longuet-Higgins, H.C., (1982) 'The role of the vertical dimension in stereoscopic vision,' *Perception,* vol. 11, pp. 377-386.

Longuet-Higgins, H.C., and Prazdny, K., (1980) 'The interpretation of a moving retinal image,' *Proc. Royal Society London,* vol. B 208, pp. 385-397.

Marr, D., (1982) **Vision,** Freeman, San Francisco.

Miles, W.R., (1931) 'Movement interpretations of the silhouette of a revolving fan,' *American J. of Psychology,* vol. 43, pp. 392-505.

Neuhaus, W., (1930) 'Experimentelle Untersuchung der Scheinbewegung,' *Arch. Ges. Psychology,* vol. 75, pp. 315-452.

Petersik, J.T., (1979) 'Three-dimensional object constancy: Coherence of a simulated rotating sphere in noise,' *Perception and Psychophysics,* vol. 25, no. 4, pp. 328-335.

Petersik, J.T., (1980) 'The effect of spatial and temporal factors on the perception of stroboscopic rotation simulations,' *Perception,* vol. 9, pp. 271-283.

Shepard, R.N., and Metzler, J., (1971) 'Mental rotation of three-dimensional objects,' *Science,* vol. 171, pp. 701-703.

Shepard, R.N., and Cooper, L.A., (1982) **Mental Images and Their Transformations,** MIT Press, Cambridge, MA.

Sperling, G., Pavel, M., Cohen, Y., Landy. M.S., and Schwartz, B.J., (1983) 'Image processing in perception and cognition,' in O.J. Braddick and A.C. Sleigh (eds.), **Physical and Biological Processing of Images,** Springer-Verlag, Berlin, pp. 359-378.

Tsai, R.Y., and Huang, T.S., (1982) 'Uniqueness and estimation of three-dimensional motion parameters of rigid objects with curved surfaces,' Rep. R-921, Coordinated Science Laboratory , University of Illinois at Urbana-Champaign.

Ullman, S., (1979) **The Interpretation of Visual Motion,** MIT Press, Cambridge, MA.

Ullman, S., (1979a) 'The interpretation of structure from motion,' *Proc. Royal Society London,* vol. B 203, pp. 405-426.

Ullman, S., (1979b) 'Relaxation and constrained optimization by local processes,' *Computer Graphics and Image Processing,* vol. 9, no. 6, pp. 115-125.

Ullman, S., (1983) 'Recent computational results in the interpretation of structure from motion,' in **Human and Machine Vision,** J. Beck, B. Hope and A. Rosenfeld (eds.), Academic Press, New York.

Ullman, S., and Hildreth, E.C., (1983) 'The measurement of visual motion,' in O.J. Braddick and A.C. Sleigh (eds.), **Physical and Biological Processing of Images,** Springer-Verlag, Berlin.

Wallach, H., and O'Connell, D.N., (1953) 'The kinetic depth effect,' *J. of Experimental Psychology,* vol. 45, pp. 205-217.

Wallach, H., O'Connell, D.N., and Neisser, U., (1953) 'The memory effect of visual perception of three-dimensional form,' *J. of Experimental Psychology,* vol 45, pp. 360-368.

Wallach, H., Weisz, A., and Adams, P. A., (1956) 'Circles and derived figures in rotation,' *American J. of Psychology,* vol. 69, pp. 48-59.

Webb, J.A., and Aggarwal, J.K., (1981) 'Visually interpreting the motions of objects in space,' *Computer,* vol. 14, no. 8, pp. 40-49.

White, B.W., and Mueser, G.E., (1960) 'Accuracy in reconstructing the arrangement of elements generating kinetic depth displays,' *J. Experimental Psychology,* vol. 60, no. 1, pp. 1-11.

CHAPTER 8

Some Problems with Correspondence

Michael Jenkin
Paul A. Kolers

8.1 Introduction

The notion of correspondence underlies many current theories of human and machine visual information processing. Algorithms for both the *correspondence process* and solutions to the **correspondence problem** have appeared regularly in the computer vision literature. Algorithms for stereopsis (Marr and Poggio, 1977; Barnard and Thompson, 1980; Mayhew and Frisby, 1980) and for tracking objects through time (Moravec, 1977; Ullman, 1979; Dreschler and Nagel, 1981; Webb, 1981; Jain and Sethi, 1984) have been presented which assume that token matching of separated or successive views is the underlying visual process. This paper will address the notion of token matching as a primitive operation in vision. We will argue that correspondence seems ill suited to the task of accounting for how an object is positioned in time or space, and that some other mechanism may provide a more apt account.

The notion of correspondence assumes that if there are two stimuli in physical space that excite two receptor organs, or if there are two receptor organs for the stimuli but only one perceptual experience, then representations of the two stimuli must come into correspondence in order to be treated by the nervous system as one object or event. This notion of correspondence is seen in Julesz's discussion of the cyclopean representation of separated retinal images (Julesz, 1971), in the concept of depth derived from stereo vision as discussed by Poggio and his colleagues (Marr and Poggio, 1977; Mayhew and Frisby, 1980), and in a treatment of apparent motion (Ullman, 1979).

For stereo vision, the assumption is made that the two images are presented simultaneously, and that the retinal images projected onto a person's two eyes separately are brought into correspondence at some later point in the visual system. These corresponding points, lines, or edges constitute the object of perception. For apparent motion, in contrast, the assumption is made that a first flash has a correspondent second flash to which it can match, and that movement is derived from the disparity of place of an object seen in one location at one moment and in another location subsequently. Kolers and Pomerantz (1971) showed that motion is seen for stimulus pairs of dissimilar shapes equally well as for pairs of similar shapes. Therefore, it is not feasible for correspondence to be drawn soley from shape. The notion of tokens was devised (Marr, 1982) to accommodate the fact that motion can be seen between different shapes, or shapes that seem to change during movement, for example, a circle seeming to become a square. A token is an abstract representation that is assumed to hold a place in perceptual experience for the corresponding shape to occupy.

Notions of correspondence underlying the perception of depth or motion assume that figure or shape is the principal coin of visual perception, i.e., that the visual system relates its experience to extracted shape information principally. Although most experiments are done with fixed shapes, thereby allowing shape to have an exorbitant role in perception research, shape is only one of many features of objects in the natural environment. If a person moves about, changing their appearance, an observer will see the same person in movement. If an observer moves about, changing their perception of the environment, the environment and the observer maintain their perceptual identity for the observer. In the laboratory, using fixed shapes as stimuli, it is easy to confuse the perceptual identity of an object with its shape, but in the natural world, objects retain their perceptual identity despite great changes in shape. That being the case, it would seem to be an inefficient technique for the visual system to attempt to match shapes in order to attain a definition or demarcation of objects. Actually, stability of the environment despite the radical sampling created by movements of the eyes may constitute a serious challenge to the notion of correspondence and of place tokens which might be used by correspondence processes. Consider, for example, that the eye in looking at a scene sweeps over a fairly large angular extent, and that central vision is restricted in extent to about 2° of visual angle. A person looking at a picture of 15° × 30° in size will make a number of fixations on different parts of the picture, but in no fixed order. The eye looks now here, now there, at the scene in a sequence of fixations, but the perceptual representation of the scene preserves scene structure rather than image sequences (Kolers, 1976). It is hard to construe how place tokens or the related notion of places that are filled with objects could be made to accommodate such sampling of an unknown scene.

We will argue in two ways that the notion of correspondence seems ill suited to the task of accounting for how an object is positioned in time or space. First, from a review of current computer vision literature we will argue that computer algorithms based on simple token matching have not been successful in most applications and that as a consequence their authors have been forced to define increasingly more sophisticated and elaborate token models and selection criteria in order to accommodate the visual information. The general failure of these algorithms to achieve object recognition, and the continual increase in the complexity of the token models and matching criteria indicate that it may not in fact be the specifics of the algorithm that are at fault, but rather that the entire methodology of token matching may be inadequate. As a second line of argument, we will present an experiment on apparent motion that suggests that the human visual system processes monocular information without regard for the token matching that is at the heart of the belief in a correspondence process, and that questions approaches that separate the structuring process from the correspondence process.

We start by defining the *correspondence problem,* and how it is thought to relate to the task of visual information processing. The issue is directly traceable to Ullman (1979), who broke the problem of interpreting visual motion into two parts, the correspondence problem and the three-dimensional interpretation problem, with the correspondence problem defined to be

> "that of identifying a portion of the changing visual array as representing a single object in motion or change" (Ullman, 1979, pp. 4).

Once a two-dimensional token correspondence has been determined, a three-dimensional interpretation task remains. This interpretation or structure from motion task takes as input a set of two-dimensional correspondences, and infers the three-dimensional structure that gave rise to the observed motion. In addition, the interpretation must decompose the changing scene into objects. Correspondence and interpretation are assumed to be totally separate:

> "Apparently, these two problems [the correspondence problem and the structure from motion problem] are solved independently by the human visual system... The critical empirical evidence for this is that none of the measurements on which the correspondence process rests involve three-dimensional angles or distances--they are all two-dimensional measurements made on the images... Thus, there is no deep need for any feedback from the later task to the earlier" (Marr, 1982, pp. 184).

Neither correspondence nor the task of recovering structure from motion has proved to be an easy problem to solve. Consider apparent motion.

When two properly placed and properly timed flashes are presented to a human observer, the person usually sees a single object moving from the first location through space to the second location. This *apparent motion* of an object has many applications in signs and as directional indicators, and has been studied intensively as a perceptual process by psychologists since Exner (1875). Kolers (1972) provides a review.

A large body of work on apparent motion has been presented as both support for the correspondence process and also as evidence of particular properties that a correspondence process should exhibit. Ullman (1979) considered apparent motion as a convenient test bed for studying the correspondence process. Indeed many correspondence algorithms claim to account for some specific examples or cases of apparent motion. Unfortunately apparent motion is, in general, more complicated than is indicated by the selected experiments that Ullman cites in his defense of a correspondence process. For example, when the two presentations are properly timed and placed but are disparate in shape and color, a small object can be seen as smoothly crossing the intervening space, making whatever geometric transformations are required for the first presentation to become the second. Squares can be seen to round out to become circles, triangles can be seen to develop an additional corner to become squares, etc., however, if the color of the first presentation is different from that of the second, the object perceived to be in motion will not be seen as changing color smoothly. If the first presentation is a red circle and the second is a green square, the single moving object will be perceived to smoothly change shape, yet abruptly change color from red to green. Such an occurrence is inconsistent with a proposal for correspondence as Ullman and others have advocated. A more elaborate constructive process seems to be at work.

Moreover, the bi-dimensionality of the correspondence process does not seem to apply to apparent motion. The human visual system can create tri-dimensionality between two figures if a third figure is interposed in the path of the apparent motion. The moving figure appears to "pop out" of the plane to pass over (or under) the intermediate figure (Kolers, 1972). We will show in an experiment presented below that human perception takes three-dimensional structure into account when resolving the disparities between apparent motion presentations. This demonstration is in conflict with the views of Marr and others.

In formulating an algorithm for the correspondence process, the type and dimension of tokens or features that must be extracted from the raw image are critical, for the correspondence process is thought not to act on the raw image itself. Elements in the raw input images lack the structure and consistency required by grouping or matching processes, and some form of preliminary process is required before a correspondence process can be applied. It has been shown that although correspondence could not take place based on the raw image data, neither could it take place based on a sophisticated representation of form (Ullman, 1979). The conclusion that there existed an intermediate representation, one not based on a complete representation of the objects in the scene, seemed inescapable. Given this unspecified intermediate representation it was claimed that correspondence must be some simple token matching process:

> "the correspondence process is low-level and autonomous and that it establishes matches between elementary constituents of images on the basis of a built-in affinity measure and local interactions" (Ullman, 1979, pp. 8).

A number of critical properties have been introduced by concerning the correspondence process (Ullman, 1979).

1) Correspondence takes place on a set of tokens extracted from the raw image data.

2) Correspondence is a simple token matching process.

3) Correspondence is established without interpretation of the three-dimensional structure of the objects being observed, i.e., correspondence is a two-dimensional task.

4) Correspondence takes place based on the primitive tokens without considering possible correspondences that could operate over higher order groups of the tokens.

Our primary argument will be with the token matching process that has become synonymous with the idea of correspondence, and thereby has become the basis of a wide range of computer vision algorithms. In addition, we will argue that regardless of how the correspondence process operates, the separation of this process from its three-dimensional interpretation appears incorrect, and that the independence of token matching from the possible matching of higher order groupings is also not generally supported by results derived from the human visual system.

8.2 Determining Correspondence

Correspondence is usually interpreted as the construction of a mapping for two sets of inputs based on a minimization of some function over all possible matches. In order to develop a correspondence algorithm two questions should be considered, 1) What are the primitives to be matched? 2) How should similarity of the primitives be measured? With these questions in mind, the solution to the correspondence problem has been formulated as follows: "The obvious answer is, take the solution which maximizes the overall similarity between the frames. This similarity can be measured by means of a standard cost function that gives a similarity value to each pairing in a given solution, the overall similarity being the sum of the values of each pairing. The cost function tells us roughly how many quite poor pairings should be accepted in order to acquire an excellent one in the overall match" (Marr, 1982, pp. 186).

This independence of matches reduces the achievement of correspondence to a simple linear problem. However, this independence is not found in general in the human visual system (Marr, 1982), and the assumption of independence has had limited success in computer vision systems. Unfortunately for computer vision, the generalization of the correspondence task to non-independent matching functions leads to expensive computational tasks which lack the ease of solution associated with the linear minimization task of the independent case.

8.3 Correspondence in Computer Vision

8.3.1 Correspondence in Stereopsis Algorithms

The correspondence process has been put forward in computer vision as a solution to stereopsis, as well as to the task of tracking objects through time. Correspondence has long been considered the *hard* problem in designing a stereopsis algorithm. Early computational models of stereopsis, e.g., (Marr and Poggio, 1976), explicitly described stereopsis as the construction of simple point-to-point matches. The inability of these algorithms to deal with stimuli other than random dot stereograms has led in more recent algorithms to the use of sophisticated matching primitives such as edges and peaks. Correspondence, however, remains as the underlying task of the computational problem of stereopsis as indicated in the following:

> "The main problem that human stereo vision has to solve is what is called the correspondence problem - how to find corresponding points in the two images without recognizing objects or their parts" (Poggio and Poggio, 1984, pp. 381).

The notion of matching zero-crossings extracted from spatial frequency tuned channels was introduced as the primitive operation in stereopsis (Marr and Poggio, 1977), while others have suggested that zero-crossings alone are not a sufficiently rich set of primitives for stereopsis (Mayhew and Frisby, 1980). Although the suggested scheme defines the monocular primitives differently, and utilizes a more sophisticated set of matching rules, the fundamental task the algorithm is required to solve is

> "How does the stereopsis system decide which point in one eye's view matches with which point in the other eye's view, and how does it resolve ambiguous or false matches?" (Mayhew, 1983, pp. 2).

The failure of the correspondence task to solve the stereo problem has actually been noted as follows: "In the case of stereopsis, the classic figural representation of the correspondence problem directly, and I think misleadingly, suggests the structure for its solution, that is facilitation between same disparity planes coupled with inhibition between different disparity planes. Whatever other value the exploration of stability in networks of excitatory and inhibitory processors might have, and 'stereopsis' may be an ideal 'preparation' for this endeavor, it's not clear that such work has greatly illuminated the nature of the stereo problem" (Mayhew, 1983, pp. 3).

As the complexity and dimensionality of the correspondence tokens increase, the complexity of the associated matching functions increases also. Current stereopsis algorithms, e.g., (Mayhew and Frisby, 1980), utilize as matching primitives edges and peaks detected over various scales of spatial frequency. The task of matching these tokens, and in particular, the task of interpreting information across different spatial frequencies, is particularly difficult.

The task of combining information from different spatial frequency tuned channels has been considered in terms of a multi-level cooperative process with possible methods of incorporating other disparity information into a multi-resolution representation (Terzopoulos, 1982). Flexible thin plate assumptions (Terzopoulos, 1982) and feed-back mechanisms (Cohen and Grossberg, 1984) suggest the complexity of the problem.

In current practice, an algorithm may be able to interpret the matches of the primitives correctly, but it is usually unable to infer or project any information concerning the intervening structure between the matches. This task of filling in the regions between known corresponding features, i.e., features of known disparity, is particularly difficult for machine approaches to vision. Thus, a fundamental question is put forward by Cohen and Grossberg (1984, pp. 120):

"Why does the nervous system bother to compute edges if it is so hard to spontaneously and unambiguously fill-in between the edges?"

Or to rephrase the problem in terms of machine vision: why are stereopsis algorithms based on the detection of correspondences between discrete artificial structures, such as edges, when by doing so the remaining problems become difficult and perhaps impossible to solve? Note that Kolers (1983) suggested that processing of edges and processing of the interiors that they bound were actually carried out by two different sorts of visual operation.

A number of computer algorithms have been developed for stereopsis. Although the choice of monocular primitives, i.e., matching primitives, has varied widely and the methods of matching have become more and more sophisticated, the underlying computational task has remained unchanged. A match for each feature in the left eye's view must be found in the right eye's view. The algorithms have had some limited success, but they have yet to approach the performance of the human visual system. Moreover, higher level processes such as cooperative global stereopsis or multi-level reconstruction are required by these correspondence algorithms in order to solve some of the difficult matching problems that arise.

8.3.2 Correspondence in Temporal Matching Algorithms

Correspondence has also been considered as the fundamental task in tracking objects through time (Ullman, 1979). The assumption of discrete presentations that underlies any correspondence process is not true, in general, when correspondence is applied to temporal matching. Only in restricted cases, such as computer vision, is the assumption of discrete presentation satisfied. It is not a usual property of the human visual system, even in the restricted case of apparent motion. In addition, any temporal correspondence process will only determine matches between presentations. The task of determining the *path* that an object follows between the presentations remains. This path problem is analogous to the filling-in problem encountered by correspondence processes when they are applied to stereopsis.

Ullman (1979) argues that given separate physical locations for two presentations, the temporally continuous extension of the correspondence problem can be reduced to the discrete case. Although his argument has merit from a computational standpoint, it does raise a number of important problems when applied to human visual processing. For the independent units suggested by Ullman to operate in this continuous fashion, a central clock and a fixed unit of time are required. There is little support for either of these notions in the human perceptual system (Kolers, 1968; Kolers and Brewster, 1985). Furthermore, it is not clear how a mechanism that relies on a temporal ordering of inputs, i.e., that event A occurs before event B, can operate under the action of eye movements (Kolers, 1976).

Many of the algorithms presented in the literature for object tracking fall within the framework of token matching presented by Ullman. The algorithms developed differ only in the complexity and dimensionality of the matching primitives chosen, as well as the type of similarity or affinity measure used. The algorithms rely on correspondence as the underlying visual process, and assume that this process is a two-dimensional task operating upon discrete presentations. Many different types of algorithm have been devised to solve the correspondence problem. One approach is to use the structure from motion problem to help solve the correspondence problem. By restricting the type of moving object seen, that it be rigid, for example, only certain combinations of motion will be valid. A typical approach of this type is that of Hoffman and Flinchbaugh (1980). They utilized a planarity assumption to restrict the types of motion that objects were allowed to undergo. They based this restriction on the observation that objects tend to move in single planes for extended periods of time. The algorithm is fairly straightforward: consider all possible correspondences and accept only those combinations that will be acceptable to the structure from motion algorithm. By carefully choosing the type of motion permitted, and the number and type of cameras with which to record, it is possible to achieve correct results in a restricted range of conditions.

Another approach is to uses a similarity of motion assumption based on smoothness of motion (Rashid, 1980; Jenkin, 1983; Jain and Sethi, 1984). In such an approach correspondences are made that satisfy a minimum cost likelihood function that measures the motion exhibited by the objects being tracked. The motion of each object is treated independently. These algorithms are based on the principle of visual inertia: objects that have a given direction of motion and speed at one instant are likely to move with the same direction and speed over the next interval (Ramachandran and Anstis, 1983). As each feature can be treated independently, these approaches are not restricted to the analysis of specific classes of objects, rather it is the class of possible motions that is restricted.

Algorithms that determine a temporal correspondence that is most consistent with a theory of common fate have been used for many years (Barnard and Thompson, 1980; Dreschler and Nagel, 1981). The assumption made is that points or objects tend to move in a manner similar to other points or objects that lie in a local neighborhood. The matching process is interpreted as a relaxation labelling task over the possible vector velocities of the matching primitives chosen. Features are usually extracted that differ from their surroundings, such as corner points (Nagel, 1982) or points of high variance (Moravec, 1977). Except for specific applications in which the type of feature to extract is known *a priori,* the task of choosing a feature is particularly difficult. [See Thorpe (1984) for a survey of current interest operators]. In addition, generalized feature points, such as those detected by the Moravec operator, are usually unstable. Once these points are located in the two images, they are matched, usually using some form of relaxation labelling (Barnard and Thompson, 1980).

Relaxation labelling is an iterative process that assigns and adjusts "probabilities" for each match between the two frames presented. The weighting formula is complex, but basically the "probability" of a given match is updated by a weighted sum of neighboring points whose matches are consistent, i.e., having nearly the same velocity. This iteration is continued until the "probabilities" stabilize. The assumption that neighboring points have similar velocities may hold for points selected from the surface of the same object provided that the object is undergoing a simple translation, but it will not hold for objects that are rotating or for scenes in which objects move at different velocities.

Assumptions such as "common fate" violate the independence of matching, a fundamental assumption of Ullman (1979). Such algorithms must search a large space without the linearity property that is inherent in Ullman's model. In order to make the search computationally feasible, heuristics are used to limit the size of the search space. These heuristics, although they may in general decrease the search time required to find the correct set of matches, may cause the correspondence process to select a local minimum, or to search aimlessly. In general, the problem is much less well behaved than the independent matching assumption suggests is the case.

The correspondence of low-level tokens, determined by finding a matching between primitives of some type and dimensionality, is the major task of many computer vision algorithms. Whether the task be stereopsis or object tracking, the approach of minimizing some function over the possible pairing of inputs has been a common theme of many algorithms. The failure of this approach is most notable in the ever increasing complexity of the tokens used by these algorithms, e.g., zero-crossings (Marr and Poggio, 1976), zero-crossing and peaks (Mayhew and Frisby, 1980), interestingness operator results (Moravec, 1977), corner operator results (Nagel, 1982), etc., and the increasing complexity of the cost function to be minimized, e.g., from simple linear minimization (Ullman, 1979) to complex relaxation labelling (Barnard and Thompson, 1980). The lack of success of these approaches may lie not in their choice of matching primitives, or in the rules used to combine the primitives, but rather it may lie in the approach itself: the assumption of matching as the fundamental operation in vision.

To illustrate this point even further, we describe a simple experiment whose results challenge the notion of correspondence applied to human perception of motion.

8.4 An Experiment on Correspondence

Apparent motion is customarily demonstrated in the two-dimensional plane by varying the timing of two otherwise similar presentations, A and B, that are at different locations. The sequence requires brief presentation of A and B and brief pauses between them. In sum, the experimental stimulus is a sequence, such as A-P1-B-P2-A..., where P1 and P2 are the brief pauses. When the timing is properly adjusted, most people see a single object oscillating smoothly between the locations of A and B. Motion is not restricted to the two-dimensional plane, however, nor need the alternated shapes be greatly similar in appearance. Several procedures based on variations of timing or shape, or on the interposition of objects in the path of motion have been described that create a perception of motion in three-dimensions (Kolers, 1972). Still another procedure, which we exploited in the present experiment, is based on variations in the size of the stimuli. Suppose that A and B are two plane figures, such as circles, squares or triangles, that differ only in diameter, and suppose that they are presented as superposed, concentric pairs. Apparent motion can then be seen between the two in either of two ways: in the two-dimensional plane as a shape that seems to shrink and expand, or in the third dimension as an object that seems to approach and recede. We exploited this illusory perception to study some aspects of depth and motion.

In one test, the stimuli were circles presented in a tachistoscope. Three shapes made up a test set, a small, medium, and large circle, where medium means an inner diameter of 1.08° visual angle, and outer diameter of 1.22°, and small and large differed by half a degree from the medium circle. The viewing distance was 50 inches (125 cm). In one test, A and B were the medium and the large circle, respectively. In another test, A and B were the medium and the small circle, respectively, as illustrated in Figure 1. Alternation of medium and small circles yielded a perception of a circle receding from and approaching a midpoint, while alternation of medium and large circle yielded a perception of a circle approaching from and then receding to the midpoint. The observer's task was to scale the depth in the display, using a procedure of numerical magnitude estimation (Gescheider, 1984). Thus, on a scale ranging from 0 to 5, the observer assigned a value of 0 or 1 if there were very little depth in the appearance of a cycled pair of shapes, and assigned a value of 5 if the apparent depth were at a maximum. The shapes were presented five times at each of three different durations and each of seven different values of pause. The observer made one judgement of depth after each cycle of five presentations. The durations were 50, 160, and 250 millisecond; the pause ranged from 20 to 300 millisecond. Brightness was about 4 millilamberts.

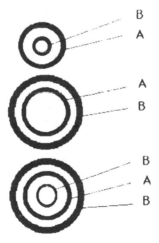

Figure 1. Superimposed circles: A and B indicate first and second presentation. The first presentation was a medium sized circle approximately 1° of visual angle.

In a third set of observations, presentation A was the medium circle, while presentation B contained both the large and the small circle, simultaneously. The question here was whether shapes could split and create two different directions of motion in depth simultaneously: receding from the medium circle to the small one, and approaching from the medium circle to the large one. The bi-directional motion perception would be analogous to the way an object can be seen to split and create two different directions of motion in the two-dimensional plane (Kolers, 1972, pp. 61 ff). The appearance of split depth in the present circumstances would be an accordion-like motion, rotated 90° to the observer, of the object simultaneously moving inward and outward from the medium circle. The sense can be achieved easily with real motion of one's hands before one's face, one hand approaching the face while the other simultaneously recedes.

When the stimuli were medium to large or medium to small, singly, the likelihood of reporting motion in depth increased with an increase in stimulus duration. The likelihood of seeing motion in depth was an inverted U-shaped function of the pause duration: low likelihood at very short and very long durations of pause and greater at in-between values (Kolers and Pomerantz, 1971). Overall, slightly more depth was reported in the direction medium to large than in the direction medium to small, 5 compared to 4.5, respectively. Thus, motion was seen in depth about equally easily receding or approaching from a midpoint.

The results were quite different when the medium circle was alternated with a larger and a smaller circle presented simultaneously. The average scaled depth then was significantly less compared to the two conditions individually. Moreover the observer did not see movement in recession and approach simultaneously, rather, the observer characteristically reported that the depth perception was between two of the shapes, not between all three. The two usually were the medium and large circles while the small circle seemed to remain in the same plane as the medium circle blinking on and off.

The import of these results is that the visual system seemed to select among the shapes in creating a sense of depth. The visual system was not able to create a sense of shapes approaching and receding simultaneously. The possibility of tri-dimensional representation seemed to be but one of many contingent options present to the visual system. Smooth changes of shape are accomplished easily when the figures are simple or without much texture. Smooth change is not accomplished when the figures contain a great deal of inner texture that the system must analyze and represent (Kolers, 1972, pp. 108 ff). In this sense, some perceptions seem to be easier for the system to accomplish than others, as plastic deformation of shapes seems to occur with shorter presentations or shorter pauses than does rotation of an object in perceptual space (Kolers, 1972).

As a check on our procedure we repeated the experiment with a modified design. An array processor created two superposed concentric squares on the screen of a color monitor. The squares measured 4.2 cm, 6.1cm, or 7.9 cm per side and the observer sat at a distance of 210 cm from the screen, making the squares 1.15°, 1.66°, or 2.15° of visual angle, respectively. Adjacent to the squares the system presented two colored lines, one red the other green. The presentations were anaglyphs, and the observers wore a pair of colored spectacles, one eyepiece red and the other green. A red line seen by one eye and a green line seen by the other were separated various distances on the screen of the CRT. By fusing the lines visually, the observer attained a perception of the lines at varied distances from the plane of the screen of the CRT. The squares appeared as yellow objects on the screen and so were seen by both eyes. The observer's task was to say whether the line appeared to be farther from the screen in depth than one square was from the other. In sum, the spatial disparity of the lines when fused created a sense of depth and the observer compared the extent of this depth relation to the depth relation of the sequentially presented squares.

Ten subjects were tested, some with some without previous experience in psychophysical observation. Four values of duration were crossed with three values of pause to create 12 stimulus combinations. Each was presented for 5 cycles to an observer, who made 1 judgement for the set of 5. A random order generator governed the sequence of combinations. Each observer went through the sequence four times, in the order A, B, B', A', where B' and A' are the reverse order of B and A. Thus each observer made four judgements of each combinations, but the judgements were based on five presentations each.

The data for the 10 subjects are not uniform. Three subjects who had extensive prior practice yielded regular curves with limited variability. Another four subjects had trends similar to the first three but with greater variability. The curves for the three experimental subjects are shown in Figure 2, while the curves for the remaining seven subjects are shown in Figure 3. It is clear that the greater variability of the second group does little to the basic relation. Figure 4 shows the data for all 10 subjects. Looked at in reverse sequence, the curves show that the more practiced observers were less likely to assign a depth relation to the combination of a medium square followed simultaneously by a larger square and a smaller square. The general tendency is for the depth relation to increase with an increase in duration of exposure and with pause; for approximately equal depth response with larger and with smaller squares; and for a decreased likelihood of depth response in the split condition. Moreover, as with the first experiment, movement in the split condition was characteristically seen between the medium and larger squares, while the smallest square blinked on and off in the same plane as the medium square.

Thus, using two different procedures, we have found that shapes do not split in the depth dimension analogous to the way that they split in the two-dimensional plane. If depth were due merely to tracking or to inferring structure from motion, then a depth relation should have occurred in both directions simultaneously. We also point out that the failure of this simultaneous depth relation to occur in apparent motion despite its ready

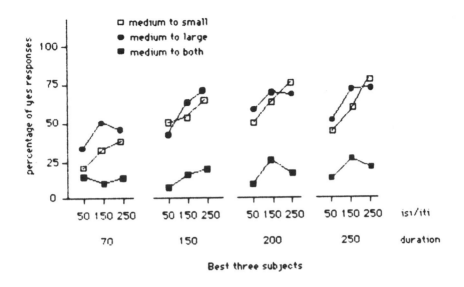

Figure 2. Likelihood of seeing motion in depth; best three subjects.

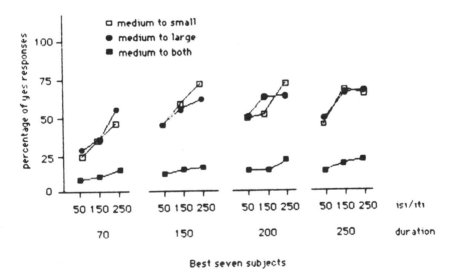

Figure 3. Likelihood of seeing motion in depth; seven subjects.

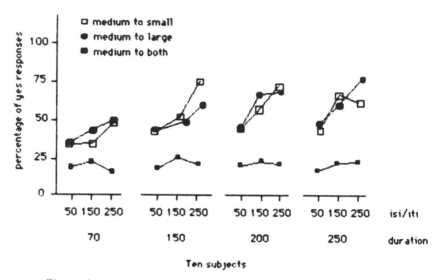

Figure 4. Likelihood of seeing motion in depth; ten subjects.

occurrence with real motion can be taken as still another way in which the visual system distinguishes its processing of the two kinds of motion. There seems to be little warrant in the behavioral data for assuming that real motion and apparent motion are due to fundamentally similar mechanisms, as some authors do. There seems to be even less warrant to assume that correspondence is an important component of processing of apparent motion. There is, after all, no problem in the visual system's finding correspondents between a first and second presentation with the displays we have used. That the large and the small square were presented simultaneously suggests that something other than correspondence is required in successful occurrences of apparent motion.

8.5 Conclusions

Ever since its introduction, constructing a correspondence between discrete presentations has become one of the underlying tasks of computer vision. The idea of reducing a complex poorly defined problem, i.e., image understanding, to a well defined one in graph theory is appealing. However, as more and more sophisticated matching rules and primitives are defined, the simple problem in graph theory to which Ullman proposed to reduce correspondence has become an intractable problem for nonlinear relaxation labelling. As computer vision researchers turn to human visual information processing for guidance, tasks such as correspondence by token matching must be reconsidered. A number of difficulties have emerged when investigators attempt to apply discrete token matching as the fundamental task in human visual processing.

(i)   A correspondence process assumes that presentations are discrete. This is not the case in the human visual system. The human visual system operates in a continuous environment. Even in the restricted case of apparent motion, there is a continuous presentation to the visual system although its stimulus may change abruptly.

(ii)   Correspondence is assumed to be a two-dimensional task that is separate from the three-dimensional structure from motion process. Results from the experiments presented above suggest that this assumption may be in error. Three-dimensional interpretation can influence the nature of any correspondences that are made.

(iii)   Correspondence tasks do not determine the path an object makes between presentations. Yet the human visual system does perform this path extrapolation. This path is extrapolated in many dimensions, e.g., position, shape, color, etc.. The type of extrapolation differs between the various dimensions. Color appears to change abruptly, whereas shape often changes smoothly, though shapes, too, can be made to change abruptly (Kolers, 1972, pp. 108 ff). Positional information is even more difficult to categorize. Paths are not necessarily straight, nor do they necessarily lie in the two-dimensional plane. Objects may appear to approach or recede, or they may even appear to jump out of and back into the plane.

These are difficult problems for the notion of correspondence to overcome. Rather than attempt to propose even more sophisticated matching rules and postulate other processes to perform the extrapolation task between discrete presentations, it may be more fruitful for investigators to consider alternate methodologies for visual information processing. Correspondence implies a detection model. Discrete presentations are compared and matches detected. This comparison and detection results in the need for a process to construct the motion of objects between the presentations or the structure of surfaces between matched features in stereopsis. If a constructive process is required for this filling-in, it is reasonable to ask if a constructive process might not be a more reasonable step for correspondence than simple token matching.

We suggest that the token matching scheme that was based on the notion of correspondence of primitives extracted from discrete presentations has failed. An alternative process, based on continuous presentations and taking into account possible three-dimensional interpretations may provide a more successful approach to the problem. A number of properties that such a mechanism should exhibit have emerged.

(1)   Correspondence should not be considered to be wholly a two-dimensional task. When the stimuli are appropriate, the human visual system perceives objects that appear to jump over other objects, moreover, correspondences can be influenced by three-dimensional interpretations of stimuli. If we generalize correspondence as defined by Ullman (1979) from a two- to three-dimensional task, and if this change is to be incorporated from the very earliest stages of visual information processing, then the separation of correspondence from a structure from motion stage seems less than justified.

(2)   Correspondence does not take place based on discrete snapshots. Presentations may change abruptly, but the visual system processes information in a continuous fashion. As shown in many experiments, and in the experiment presented above, the timing of stimuli is critical for the type of interpretation formed by the visual system.

(3)   Correspondence can be built up by repetition. It may take a number of repetitions of a set of stimuli before an interpretation can be formed. In addition, different earlier presentations may influence the interpretation of a later one. Correspondence possesses a memory.

(4)   Correspondence must not only determine where an object went, it must also determine how it got there. This includes not only extrapolation of path, which may be three-dimensional, it also includes extrapolation of form and color. In stereopsis applications, correspondence must infer the surfaces, (possibly illusory) between features.

In this chapter we have discussed correspondence as the underlying task in a number of visual processes. We have examined the notion of correspondence, both as a paradigm for machine models of visual processing as well as for human visual processing. Although some success has been found with correspondence as a model for machine vision, the successes have been limited. Correspondence is quite unable to account for a number of the properties of human perception.

We have presented an experiment and reviewed previous psychological results which contradict the notion of the separation of correspondence as a two-dimensional process from structure from motion as a three-dimensional process.

Finally, we have suggested that if token matching is unable to perform the required visual tasks, then a different approach should be considered. We have proposed such an approach, one based on a constructive task rather than on the detection of similarities of shapes.

References

Barnard, S. and Thompson, W., (1980) 'Disparity analysis of images,' *IEEE Transactions on Pattern Analysis and Machine Intelligence,* PAMI-2, no. 4.

Cohen, M.A., and Grossberg, S., (1984) 'Some global properties of binocular resonances: Disparity matching, filling-in, and figure-ground synthesis,' **Figural Synthesis**, P. Dodwell and T. Caelli (eds.), Lawrence Erlbaum Associates.

Dreschler, L., and Nagel, H.-H., (1981) 'On the frame-to-frame correspondence between greyvalue characteristics in the images of moving objects,' *Proc. Int. Joint Conf. on Artificial Intelligence,* Vancouver, Canada.

Exner, S., (1875) 'Ueber das sehen von bewegungen und die theorie des zusammengesetzen auges, *Sitzungsberichte Akademie Wissenchaft,* vol. 72, pp. 156-190.

Gescheider, G.A., (1984) **Psychophysics: Method, Theory, and Application**, Erlbaum, Hillsdale, NJ.

Hoffman, D., and Flinchbaugh, B., (1980) 'The interpretation of biological motion,' AI Memo no. 608, MIT.

Jain, R., and Sethi, I., (1984) 'Establishing correspondence of non-rigid objects using smoothness of motion,' *Proc. 2nd IEEE Workshop on Computer Vision: Representation and Control,* Annapolis, MD, pp. 83-37.

Jenkin, M., (1983) 'Tracking three-dimensional moving light displays,' *ACM Interdisciplinary Workshop on Motion: Representation and Perception,* Toronto, Canada.

Julesz, B., (1971) **Foundations of Cyclopean Perception**, Bell Labs. Inc..

Kolers, P.A., (1968) 'Some psychological aspects of pattern recognition,' in **Recognizing Patterns**, P.A. Kolers, and M. Eden (eds.), MIT Press, Cambridge, MA.

Kolers, P.A., (1972) **Aspects of Motion Perception**, Pergamon Press Ltd., Headington Hill Hall, Oxford.

Kolers, P.A., (1976) 'Buswell's discoveries,' in **Eye Movements and Psychological Processes**, R.A. Monty, and J.W. Senders (eds.), Erlbaum, Hillsdale, NJ.

Kolers, P.A., (1983) 'Some features of visual form,' *Computer Vision, Graphics, and Image Processing*, vol. 23, pp. 15-41.

Kolers, P.A., and Brewster, J.M., (1985) 'Rhythms and responses,' *J. of Experimental Psychology: Human Perception and Performance*, vol. 11, pp. 150-167.

Kolers, P.A., and Pomerantz, J.R., (1971) 'Figural change in apparent motion,' *J. of Experimental Psychology*, vol. 87, pp. 99-108.

Marr, D, (1982) **Vision**, Freeman Press.

Marr, D., and Poggio, T., (1976) 'Co-operative computation of stereo disparity,' *Science*, vol. 194, pp. 283-287.

Marr, D., and Poggio, T., (1977) 'A theory of human stereo vision,' AI Memo, no. 451, MIT.

Mayhew, J., (1983) 'Models of stereopsis,' *Preliminary Proc. Workshop on Vision, Brain and Cooperative Computation*, University of Massachusetts.

Mayhew, J., and Frisby, J., (1980) 'The computation of binocular edges,' *Perception,* vol. 9, pp. 69-86.

Moravec, H., (1977) 'Towards automatic visual obstacle avoidance,' *Proc. Intl. Joint Conf. on Artificial Intelligence,* Cambridge, MA, p. 54.

Nagel, H.-H., (1982) 'On change detection and displacement estimation in image sequences,' *Pattern Recognition Letters,* vol. 1, pp. 55-59.

Poggio, G., and Poggio, T., (1984) 'The analysis of stereopsis,' *Annual Rev. Neuroscience,* pp. 379-412.

Ramachandran, V.S., and Anstis, S.M., (1983) 'Extrapolation of motion path in human visual perception,' *Vision Research,* vol. 23, pp. 83-85.

Rashid, R., (1980) 'Towards a system for the interpretation of moving light displays,' *IEEE Pattern Analysis and Machine Intelligence,* PAMI-2, pp. 574-581.

Terzopoulos, D., (1982) 'Multi-level reconstruction of visual surfaces: Variational principles and Finite element representations,' AI Memo, no. 671, MIT.

Thorpe, C.E., (1984) 'An analysis of interest operators for FIDO,' *Proc. 2nd IEEE Workshop on Computer Vision: Representation and Control,* Annapolis, MD.

Ullman, S., (1979) **The Interpretation of Visual Motion,** MIT Press, Cambridge, MA.

Webb, J., (1981) 'Structure from motion of rigid and jointed objects,' *Proc. Intl. Joint Conf. on Artificial Intelligence,* Vancouver, Canada.

CHAPTER 9

Recovering Connectivity from Moving Point-Light Displays

Dennis R. Proffitt
Bennett I. Bertenthal

9.1 Introduction

An important trend in the visual sciences is the emerging convergence between psychophysical and computational approaches to visual information processing (Beck, Hope, and Rosenfeld, 1983). Each field is concerned with similar issues and problems; however, each applies a sufficiently different approach to provide complementary lines of investigation. One point of convergence is found in current "bootstrapping" procedures that analyze visual information into minimal stimulus conditions and then seek to model processes by which these conditions can be transformed into relevant environmental properties.

Suppose that we are interested in the extraction of some environmental property, P, from optical information. P could be size, shape, reflectance, or any other property of distal objects. A bootstrapping approach typically involves systematic investigations of the following three questions:

1. What are the minimal stimulus conditions for perceiving P? To answer this question we decompose stimuli that typically evoke the perception of P into minimal informational sources and then evaluate

each separately. For example, computer-generated random dot stereograms have been devised in order to isolate the contribution of
stereopsis in perceiving depth relations (Julesz, 1960).

2. By what process(es) can these minimal conditions be transformed
into P? This computational question is of equal concern to psychologists and computational researchers. Once we have shown that
environmental property P can be extracted from some minimal
stimulus condition, S, we seek to discover the formal processing constraints by which S's structure is mapped onto a specification of P.
Returning to the random dot stereogram example, there are a number
of computational models for extracting edges from stereo images
(e.g., Marr and Poggio, 1979).

3. How does the human visual system transform minimal properties
into P? Typically, there are a variety of means by which stimulus
properties can be mapped onto perceptual properties. Research on
this question investigates whether the human visual system imple-
ments those processing constraints proposed by existing formal
models. If it turns out that aspects of human performance are not
correctly predicted by these models, then the bootstrapping procedure
cycles back to question 2. Although this is primarily a psychological
question, research in human visual processing may provide valuable
suggestions for computer vision research.

The present chapter traces the bootstrapping approach employed in
investigations of the recovery of form, i.e. the three-dimensional structure
of objects, from optical motion information. We discuss psychological
research demonstrating that two-dimensional motion patterns are a minimal
condition for perceiving form. Next we discuss a computer vision model
that recovers form from motion through the implementation of processing
constraints that assume local rigidity. Finally, we compare this model with
human performance. In regard to the latter, we find that aspects of human
performance are predicted by the computational model; however, people

are found to implement a number of other processing constraints that both augment and interfere with the recovery of three-dimensional structure. These other processing constraints are elaborated upon in the second half of the chapter.

9.2 Motion Information is a Minimal Stimulus Condition for the Perception of Form

One of the earliest and best known demonstrations of the perception of three-dimensional structure from transforming two-dimensional patterns was created by Wallach and O'Connell (1953/1976), who dubbed the phenomenon the "kinetic depth effect" (KDE). In their experiments, shadows of solid and wire revolving objects were cast onto a screen, and in most cases, observers perceived correct three-dimensional interpretations when viewing these rotating two-dimensional projections. Similar results were obtained when dynamic point-light projections of revolving rigid objects were presented (Green, 1961; Braunstein, 1962; Ullman, 1979; Roach and Aggarwal, 1980). Although many investigators suggested that, in interpreting such displays, the visual system might be implementing processes that assume a global rigidity constraint, it was soon recognized that people could recover three-dimensional structure from moving projections of jointed objects that do not possess global rigidity. The particular demonstration that caught the attention of motion perception researchers was a display consisting of point-lights attached to the major joints of an unseen walking person (see Figure 1). It was found that these displays provided observers with compelling impressions of locomoting people. In a later investigation (Johansson, 1976), it was found that 0.1-0.2 seconds was a sufficient exposure duration for perceiving the human body. It has been shown that friends can be recognized by their gait in point-light walker displays (Cutting and Kozlowski, 1977), and evidence has been reported of gender identification in these reduced stimuli (Cutting, Proffitt, and Kozlowski, 1978). Even infants extract a three-dimensional structure consistent with the form of a person when presented with a point-light walker display (Bertenthal, Proffitt, Spetner, and Thomas, 1985). The tractability of point-light walker displays has been greatly facilitated by the development of methods for their generation through computer synthesis (Cutting, 1978; Bertenthal and Kramer, 1984).

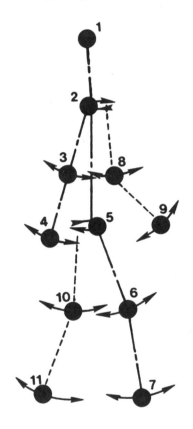

Figure 1. Eleven point-lights attached to the head and major joints of an unseen walking person.

9.3 Processing Models for Recovering Form from Motion

From a mathematical standpoint, point-light walker displays are ambiguous and do not represent any unique three-dimensional object. Johansson (1973, 1976), however, found no evidence of multistability, i.e., the perception of more than one stable interpretation in a single distal stimulus (Attneave, 1971). In seeking constraints on perceptual processing that would reduce the infinite number of possible interpretations to the single percept of appropriate biomechanical structure, Johansson proposed a perceptual vector analysis model that parsed absolute motions into relative and common motion components. This model analyzes manifest absolute motions into a common vector present in all point-lights and a set of relative motion vectors that describe the twisting and pendular movements of the body's component structures. The common motion vector specifies a perceived observer-relative displacement and serves as a frame of reference for the set of relative motions that provides information for object recognition. As a processing model, this formulation is suggestive at best, since it fails to specify the connectivity between appropriate point-lights; moreover, it fails to derive any three-dimensional interpretation, even though such an interpretation is generally reported by observers.

Proffitt and Cutting (1980a; Cutting and Proffitt, 1981) elaborated on the perceptual vector analysis model (Johansson, 1973, 1976). They proposed that the perceptual system performed a logical sequence of information extraction, beginning with relative motions as they occurred about centers of moment, i.e., the points in the arrays about which all movement appears to occur. A further constraint used in this model was a "minimum principle" applied to relative and common motion components (Cutting and Proffitt, 1982). As with the earlier approach, this model does not specify constraints that determine which point-lights should be related. Thus, the model assumes, for example, that the motions of the hip and knee point-lights are related; however, there is nothing in the model to eliminate the relating of the wrist and knee motions. The approach also fails to derive any three-dimensional interpretations.

Computer vision researchers have proposed more adequate processing models for displays of moving jointed objects, and their models take two quite different approaches. The first model operates in a top-down fashion. It takes the human form as given and seeks to match the presented figure to the known model (O'Rourke and Badler, 1980). This model lacks generality, however, because it cannot recognize any pattern other than human walkers.

The second model replies upon a "fixed-axis" assumption. It operates in a bottom-up fashion and seeks to discover the connectivity in the presented pattern by employing a set of assumptions about the motion of rigidly related points rotating in different planes of depth (Hoffman and Flinchbaugh, 1982; Webb and Aggarwal, 1982). Unlike models which seek to derive connectivity under the assumption of global rigidity (Ullman, 1979; Roach and Aggarwal, 1980), this model attempts to establish connectivity patterns under the assumption of local rigidity between pairs of points. In essence, the model seeks to recover rigid relations by testing whether each pairing of points is rigidly translating or rotating about an axis fixed in direction. Whenever pairs of points are found to meet this assumption, they are interpreted as rigidly connected. Whenever the motions of two points fail to fit the pairwise rigidity assumption, the points are interpreted as not connected. After deriving a set of pairwise connections, this model then proceeds to connect pairs having one point in common. Consider, for example, how this model finds connectivity for a point-light walker. In examining all possible pairings of the point-lights, the model will find that a knee light comprises a pairwise rigid structure with an ankle and also with a hip point-light. The hip will be found to be connected with the shoulder, the shoulder with the elbow, and so forth. Thus, the pairwise rigidity assumption represents an extremely general local processing constraint enabling the extraction of coherent, three-dimensional structure from projections of *rigid or jointed* moving objects.

There are circumstances in which fixed-axis models will fail to recover connectivity from the motions of jointed objects. These failures occur whenever the axis of rotation for a pair of points is itself rotating about a distinct axis of rotation. The models examine trajectory information for pairs of points utilizing a fixed-axis assumption: "Every rigid object movement consists of a translation plus a rotation about an axis that is fixed in direction for short periods of time" (Webb and Aggarwal, 1982, p.111). Under this constraint and parallel projection, rigidly connected pairs of points can only move relative to each other in circles that will project as circular or ellipsoidal trajectories. Any object motion that violates this constraint will produce relative motions for pairs of rigidly connected points that cannot be decomposed into circular or ellipsoidal paths. In such cases, the model will assign an unconnected status to the pair.

Figure 2 shows two situations in which the fixed-axis model will successfully recover connectivity and one in which it will fail. The top panel depicts two situations in which one point is oscillating about another. In both of these events, the plane containing the two points is fixed in its direction, referred to as the planarity constraint (Hoffman and Flinchbaugh, 1982). The second panel shows two points revolving together on a sphere. Although the motion of these two points does not define a plane, their axis of rotation maintains a constant direction and the fixed-axis model (Webb and Aggarwal, 1982) can successfully extract the relative ellipsoidal motion between the two points, thereby recovering their connectivity. The bottom panel depicts one point oscillating about another point in a plane. That plane, in addition, is revolving about an axis which is not perpendicular to the plane. Since the projected relative motion of these two points cannot be decomposed into circular or ellipsoidal trajectories, the fixed-axis model will erroneously assign an unconnected status to this pair.

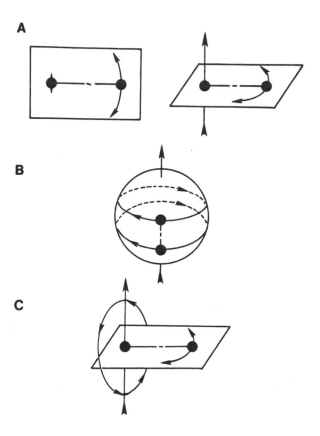

Figure 2. Fixed-axis models can recover connectivity in Panels A and
B, but not C.

9.4 Do Fixed-Axis Models Predict Human Performance?

An example of the ability of human observers to perceive connectivity in displays that violate the fixed-axis assumption is found in veridical
perception of point-light displays of twirling dancers (Mass, Johansson,
Jansson and Runeson, 1971). It has also been suggested that observers recover the connectivity pattern in these displays during periods in which the
pairwise rigid motions fall within non-rotating planes (Hoffman and
Flinchbaugh, 1982). In contrast, other models involving top-down process-

ing would be more likely to account for the preservation of these patterns during twirling transformations. Observer familiarity with the human structure makes this a reasonable proposal.

In a series of experiments we tested directly whether the human visual system could use trajectory information in a manner predicted by fixed-axis models (Proffitt and Bertenthal, 1984). We created point-light displays depicting jointed objects moving in a manner that either met or violated the fixed-axis assumption. We reasoned that if people saw rigid connectivity in the former displays but not in the latter, then the human visual system must be able to implement analog processes that seek circular and ellipsoidal trajectories in recovering form from motion.

Figure 3 depicts an example of the type of computer-generated displays used in this research. Points A and D were programmed to move as if they were affixed to a rigid rod that rotated in depth about point A. Point X was programmed to oscillate about D as D revolved. In some conditions X oscillated about D in a plane having a constant direction, and in other displays the plane containing points X and D revolved continuously. Observers saw both types of displays and judged whether points X and D appeared to be rigidly connected.

A primary finding of this research was that observers' judgments were highly consistent with a fixed-axis model. Observers judged the connection between points X and D to be rigid only when the plane containing their relative oscillation was fixed in its direction. In those cases where the fixed-axis assumption was violated, observers reported that points X and D appeared unconnected or were attached in an elastic manner, such as by a rubber band.

A second major finding of these studies was not anticipated. Identical relative motions, either circular or ellipsoidal, between pairs of points were sometimes seen as specifying connectivity, and sometimes not, depending upon the configuration and motions of other points in the display. An investigation into this matter revealed two important things

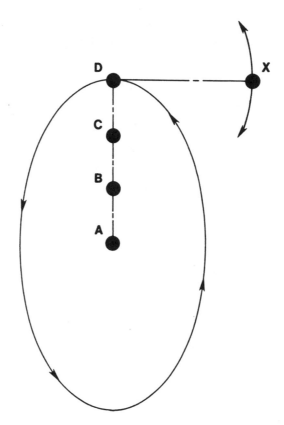

Figure 3. Points *A, B, C* and *D* are revolving in depth. Point *X* is rigidly connected to Point *D* and is oscillating around it.

about the human visual system: (1) People do not recover form from motion by testing two points at a time independent of other points in the array. Pairs of points with identical relative motions may be seen as connected or unconnected depending on the location and motions of other points in the display. (2) In extracting relative motions, the human visual system first defines a location for the rotational axis. The perceptual determination of this rotational center is discussed in the next section. The perceived location of this axis can, in some cases, interfere with the recovery of connectivity. In summary, a fixed-axis model predicts human performance in some stimulus conditions, but not others. Our research program suggests that the human visual system is implementing a number of other processing constraints in addition to those that seek pairwise rigidity. We discuss five constraints that are likely to prove useful in computer vision models.

9.5 Human Implementation of Additional Processing Constraints

9.5.1 Centers of Moment

As mentioned above, when people view rigidly connected point-lights rotating in depth, they perceive not only a connectivity pattern, but also a point of origin for the rotational motion. This point, called the center of moment (Proffitt and Cutting, 1980a), is often not marked by any of the point-lights in the array, yet its location influences which motion components are perceived. Research findings on perceiving centers of moment can be summarized graphically.

Figure 4a depicts the absolute motion paths of three points mounted on an unseen rolling wheel. These absolute motions are typically not seen; rather, the motions depicted in Figure 4b are more commonly observed. The perceptual system analyzes the absolute motions of the three points into two motion components. First, the relative motions of the points are seen as rotations of each point around a center of moment defined as the centroid of the configuration. Second, the common motion of this perceptual group is seen as the motion of the centroid. These motion components, when combined vectorally, are equivalent to the absolute motions of the individual points. The perceptual importance of the centroid is that it both serves as the center of rotations for relative motions and manifests the common motion for the configural whole.

This perceptual analysis has been found to occur for parallel projections of point-lights revolving in slanted depth planes (Proffitt and Cutting, 1979), and for revolving shapes, having curvilinear boundaries (Proffitt and

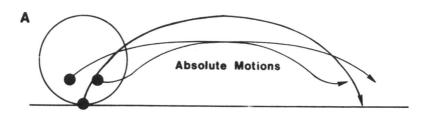

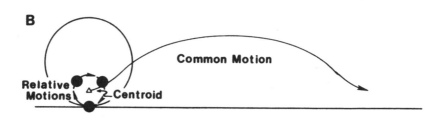

Figure 4. The absolute and perceived motions of three points on an unseen rolling wheel.

Cutting, 1980b). In examining the perceptual derivation of centers of moment for bounded shapes, it has been found that the perceptual system relies almost exclusively on boundary contour rather than luminance distribution (Proffitt, Thomas and O'Brien, 1983). A motion minimization heuristic has been proposed to describe the perceptual processes that derive perceived centers of moment (Cutting and Proffitt, 1982).

In view of the above findings, it is noteworthy that current computer vision models designed to recover connectivity patterns do not analyze the presented motions into relative and common motion components. Parsing the motions in the visual scene into these two components is of extreme utility to human observers. Relative motions define the kinematic relations within the connected figure, while the common motion defines the relationship of the figure as a whole to the observer (Proffitt and Cutting, 1980a, Cutting, 1983b). The special role of common motions in human perception is discussed in a later section. In order to analyze manifest motions into these two components, the location of the axis or axes of rotation must be determined. If one attempted to develop a computational theory that involved the extraction of relative and common motion components, it might prove useful to attempt a modeling of the perceptual system's minimization processes for the selection of centers of moment (Cutting and Proffitt, 1982).

9.5.2 Occlusion's Effect on Depth Order and Implicit Form

Computational models seeking pairwise rigidity can derive connectivity relations; however, their solutions are unique only up to reflections in depth. Thus, two potential sources of ambiguity remain after the application of the pairwise rigidity constraint. The first is illustrated in Figure 5. Figure 5a shows an image of a point-light walker, and Figures 5b, c, and d represent discrete images of four points within the walker display, selected at 90° intervals during the gait cycle. Figure 5b depicts four point-lights moving as if attached to the shoulders and elbows of the person shown in Figure 5a who appears to be walking on a treadmill. Point-lights A and X, the shoulders, oscillate back and forth, 180° out of phase. Point-light B swings with a pendular motion about A, and point-light Y moves in a similar manner about X. These latter point-lights are the elbows. Figure 5c shows the alignment, occurring twice each cycle, in which AB and XY coincide. This overlapping of one element by another leads to more than one interpretation as the elements begin again to separate. Figure 5d depicts two of the most likely interpretations of the event. To the left is shown an interpretation in which the point-light pair AB "recoils" after "contact" with the pair XY, and the latter pair likewise reverses its motion. The drawing to the right depicts the biomechanical interpretation in which the point pairs continue their original motions, presumably in different planes. A similar analysis has been provided for pairwise rigid structures on spinning disks (Martin and Aggarwal, 1981, Figure 6.2a-d). Models seeking pairwise rigidity cannot distinguish reversals in direction from continuous velocities at those points where pairwise rigid structures coincide.

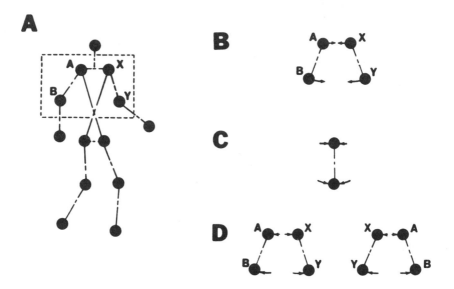

Figure 5. Without occlusion, ambiguity occurs when pairwise rigid structures coincide.

A second source of ambiguity is illustrated in Figure 6 and was acknowledged in earlier research (Hoffman and Flinchbaugh, 1982; Webb and Aggarwal, 1982). Even when an interpretation of the event shown in Figure 5 is selected in which the two rigid pairs are separated in depth and continue their motions, the order of their depth relations remains ambiguous. That is, pairwise rigidity models derive interpretations that exclude considerations of depth order. Figure 6a depicts two possibilities: The plane containing the point-lights *A* and *B* maintains a constant depth order, either in front of or behind the plane bearing *X* and *Y*. Figure 6b shows a second set of interpretations: The rigid pairs, *AB* and *XY*, maintain only a constant *depth relation* while their *depth order* is continuously changing. In this case the pairs would be seen as revolving in depth, 180° out of phase. Note that direction of rotation is ambiguous (Braunstein, 1983).

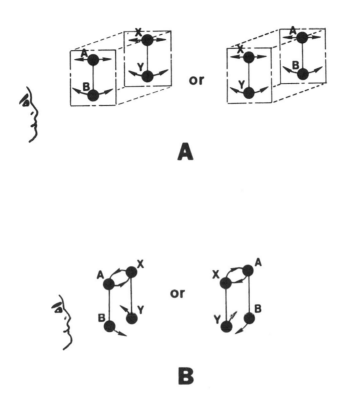

Figure 6. Panels A and B show the problem of depth order in deriving
interpretations for the event depicted in Figure 5.

This state of affairs caused us to seek additional constraints in the
distal displays that would further constrain their ambiguity, and to assess
observer sensitivity to this added information. We realized that typical
point-light walker displays contain dynamic occlusion information, a
powerful constraint not utilized in existing processing models of this
phenomenon. Since Johansson's (1973, 1976) displays were fabricated by
attaching luminance sources to real people, many of these point-lights were
successively covered and uncovered from sight by the walkers' bodies.
When, for example, a point-light walker translates along a path normal to
an observer's line of sight, the far shoulder and hip point-lights are con-

cealed entirely, and the far elbow, wrist, knee, and ankle point-lights appear and disappear as they are occluded and disoccluded by the body.

Occlusion provides a means for referencing point-lights with regard to depth order. There is considerable evidence from studies of other phenomena that people can make appropriate use of such information. Occlusion functions to define depth order under the assumption that an occluded element is always more distant from the observer than a nonoccluded element appearing at the same spatial location (Gibson, Kaplan, Reynolds, and Wheeler, 1969). Braunstein, Andersen, and Riefer (1982, Andersen and Braunstein, 1983) directly examined sensitivity to dynamic occlusion as a constraint in perceived multistability. Their display consisted of elements moving as if attached to an unseen revolving sphere. Using perceived direction of rotation as their dependent measure, they found that depth order ambiguity was greatly reduced by the introduction of occlusion.

As an auxiliary to pairwise rigidity constraints, we propose an occlusion constraint (see Figures 5 and 6): If the point-light pair *XY* is made to disappear and appear just prior to and following its coincidence with *AB*, and this appearance and disappearance is taken as representing the occlusion of *XY* by some structure lying between the planes containing *AB* and *XY*, then a unique solution is determined because the depth order of the point-lights is specified. The recoiling interpretation on the left of Figure 5c is eliminated since the continuous paths of the closest point-lights are not interrupted by a "collision" with the more distant points. The depth order ambiguity depicted in Figure 6 is eliminated, since the appearing and disappearing points are always seen as being more distant than those that are not occluded.

We examined observer sensitivity to the presence of occlusion in computer-generated point-light walker displays consisting of 11 point-lights moving as if attached to a person walking on a treadmill (Proffitt, Bertenthal, and Roberts, 1984). In one experiment, one group of observers viewed this display with no occlusion and another group viewed a display in which the appropriate point-lights blinked on and off as they would if occluded by the walker's unseen body. During the 1.5 minute viewing period, observers recorded the number of times their perception switched from one stable interpretation to another. After this viewing period, observers described what they had seen. As predicted, the display without occlusion was seen as multistable due to the depth order ambiguity. Only 33% of those who viewed this display identified it as a person walking. The presence of occlusion significantly reduced perceived ambiguity and facilitated recognition of the human form. For the second display 85% of the observers identified it as depicting a walking person.

A second study found evidence that the occlusion of point-lights is used by observers not only to specify depth order, but also to define implicit occluding contours (Proffitt, Bertenthal, and Roberts, 1984). Two different versions of the point-light walker display were used. As shown in Figure 7, one stimulus presented appropriate occlusion and the other showed occlusion that occurred in locations that are inconsistent with the implicit form of the human body. Note that depth order was equally specified in both stimuli, whereas the structure of the implicit occluding form was manipulated. Observers viewing the stimulus with inappropriate occlusion found it to be more multistable than did those who saw the canonical stimulus. Only 50% of the observers who saw the inappropriate occlusion display reported that they saw a walking person, as opposed to 85% of the observers who saw the canonical display.

The implications of this research are quite clear. Human observers take the appearance and disappearance of point-lights as specifying their occlusion by intermediate unseen forms. This information serves to define both depth order relations and the shape of the intermediate occluding surface. In this regard, it is interesting to note that sensitivity to occlusion information for specifying both of these properties emerges well within the first year of life (Bertenthal, et al., 1985), suggesting that the occlusion assumption is one of the more basic constraints used by the human visual

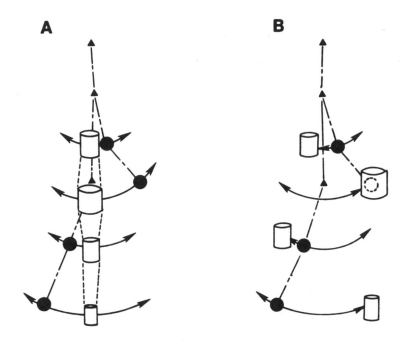

Figure 7. Cylinders show the locations where point-lights disappear for appropriate and inappropriate occlusion displays, respectively.

system.

9.5.3 Common Motion as a Grouping Factor

In seeking to recover pairwise rigidity, current computational models exploit the invariant properties of rigidly connected pairs of points moving with circular or translatory motions. These models operate locally on all possible pairings of points in the array.

There is considerable psychophysical evidence that human observers use perceived common motions as a grouping factor (Johansson, 1950; Restle, 1979). Global common motions are, by definition, manifest in every pairwise rigid structure; thus, stimulus displays presenting a common motion will be given an appropriate connectivity interpretation by pairwise rigidity models. However, pairwise rigidity models make no special use of common motions, whereas human observers do. In fact, it can be shown that the relative difficulty encountered when recovering connectivity in two different point-light arrays can be reversed depending upon whether common motion is appreciated by the system.

Consider, for example, the point-light walker display without occlusion. When presented without translation or occlusion, only 33% of those who observed this display saw it as depicting a person. When the same non-occluding display was shown with translation present, 100% of the observers identified it as a person (Proffitt, 1983). Since the absolute motions presented in the latter display are more complex than those in the first, pairwise rigidity models would encounter greater difficulty in recovering its connectivity pattern.

9.5.4 Proximity

Proximity is known to produce a strong influence on perceptual grouping processes (Wertheimer, 1923/1938). Proximity has been found to be influential in organizing perceived relative motions in moving point-light displays (Gogel, 1974; Proffitt, 1981). In addition, an effective computational rule has been proposed for deriving proximity groupings in dynamic scenes (Flinchbaugh and Chandrasekaran, 1981).

The role of proximity in perceiving connectivity in point-light arrays of jointed objects is currently under investigation in our laboratory. Our preliminary findings indicate that proximal points are more likely to be seen as connected than more distant points. In addition, proximity based connections can override connectivity patterns defined by pairwise rigidity. Figure 8 shows the motion paths of three point-lights. Point-light C is rigidly connected to point-light B. This connectivity is seen except when point-light C is much closer to point-light A than it is to point-light B. When point-lights A and C come close to each other they form a perceptual group that disrupts the connectivity seen between point-lights B and C.

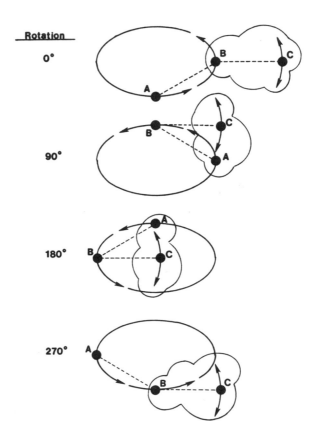

Figure 8. Dominant perceptual groupings for point-light *C* are depicted by clouds drawn around point-light pairs.

9.5.5 Familiarity

Familiarity, expectations, context, and other top-down constraints are known to have enormous influences in the recognition processes of human observers. The role of familiarity in the perception of dynamic point-light displays of jointed moving objects is difficult to assess, since the only jointed events that have been extensively examined in the psychophysical literature is the point-light walker display. Human observers are probably more familiar with the human body than they are with any other jointed object.

We sought to assess whether familiarity would influence the perception of point-light walker displays by introducing a natural context to this event (Proffitt, 1983). O'Rourke and Badler's (1980) top-down model would certainly predict an effect for familiarity since it is only implemented when there is some reason for the observer to expect that the display presented might be a person walking. In one experiment, the non-translating point-light walker display without occlusion was presented with a surface of support, depicted in perspective, beneath its feet (see Figure 9). This surface was either (1) stationary, (2) moving correctly relative to the walker's gait, or (3) moving at the same rate but in the wrong direction. Observers again viewed the display for 1.5 minutes. Results showed that the presence of the stationary surfaces improved recognition; 74% of the observers reported seeing a person when the surface was present as compared to 33% when it was absent. A moving surface greatly facilitated recognition: 100% for the appropriately moving surface and 91% for the inappropriately moving surface. Note that although the correctly moving surface yielded 100% recognition, the incorrectly moving surface yielded better recognition than the display with no surface motion at all. It is likely that the perspective structure, provided by the surface, suggested the presence of depth relations in the parallel projected point-light walker, and the motion of the surface helped to specify appropriate oscillations of the walker's component parts, thereby reducing the likelihood of seeing an inappropriate rotation interpretation.

Our program of research on infant sensitivity to biomechanical motions provides additional evidence that familiarity operates as a fundamental constraint of the visual system. In two separate series of experiments (Bertenthal, et al., 1985; Bertenthal and Proffitt, 1986), we found that infants discriminated a point-light walker display from a similar display that was systematically perturbed - either by changing the phase relations of the moving joints or by changing the phase relations of the pattern of occlu-

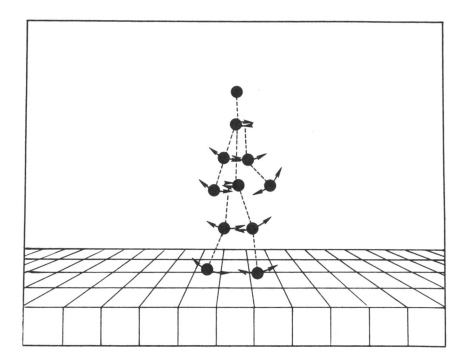

Figure 9. A nontranslating point-light walker with a surface of support beneath its feet.

sion (see Section 9.5.2). Yet, when these same displays were presented upside down, infants showed absolutely no evidence of discrimination. The most parsimonious interpretation for these findings is that the point-light walker display in a canonical or upright orientation is perceived as familiar whereas the inverted display is not. Familiarity then operates to further constrain the interpretation of the upright displays, which explains why discrimination is found only for these displays.

9.6 Incompatibilities Between Human Performance and Models Seeking Local Rigidity

Although the evidence suggests that fixed-axis models do predict human performance by either operating alone or in interaction with other constraints, these models are not always good predictors of human visual performance. In certain situations people can do things that these computational models cannot do, and vice versa. It is important to recognize that there are fundamental differences in the ways in which machines and humans process information. Machines process information and perform calculations more rapidly and more precisely, yet in so doing, they lack the flexibility of the human information processor, and furthermore, encounter greater difficulty in processing noisy or highly variable information. In this section we first discuss instances of human capabilities that exceed those of computational models, and second, we discuss some human performance limitations.

9.6.1 Human Capabilities That Exceed Fixed-axis Models: The Local Rigidity Assumption

Fixed-axis models are limited in that they seek to recover connectivity patterns in jointed objects by seeking pairwise rigid relations. Yet, there is ample evidence that human observers can recover connectivity patterns from displays presenting elastic motions. It has been shown that observers can reliably recognize emotional expressions presented in dynamic displays that consist exclusively of point-light sources that are densely distributed on the face of an unseen actor (Bassili, 1978; Spetner and Bertenthal, 1983). Another demonstration consisted of points moving as if attached to a blade of grass being blown by the wind; this was appropriately perceived as a bending form (Johansson, 1983).

Computational models that rely on rigidity assumptions to derive patterns of connectivity require that these assumptions be met almost perfectly. Small violations of the invariant properties of rotating or translating rigidly connected pairs of points will cause these models to assign an unconnected status to the pair. Our ongoing research on perceiving connectivity in unfamiliar jointed objects is providing preliminary evidence that this is not the case for human observers. Rather, very small perturbations seem not to be noticed and larger violations of the rigidity assumption cause the pair of points to be seen as connected but moving with elastic relations. In view of findings such as these, it is not surprising to find increasing recognition among computational researchers (e.g., Ullman, 1985), that current computational models must be modified if they are going to accurately model human visual information processing. We are currently extending our research to identify the important parameters and boundary conditions in perceiving elastic transformations.

9.6.2 Human Performance Limitations

Helmholtz's (1857/1962) characterization of the perceiver as an organizer and geometer has been given substance by the work in computer vision. Work in human visual perception, on the other hand, has found that many processes operate quantitatively in a rather imprecise manner, so much so that it has been suggested that the perceiver must be considered a "sloppy geometer" (Perkins, 1982). Cutting (1983a), likewise, has reminded us that mathematical formalisms are absolute, and not subject to the threshold considerations inherent to human visual processing.

The human visual system has performance limitations attendant to all of its processing capabilities. Only recently have those limitations inherent to motion information processing been examined (Llewellyn, 1971; Johnston, White, and Cumming, 1973; Regan and Beverley, 1982; Doner, Lappin, and Perfetto, 1984). Little is known about the performance limitations on perceiving jointed objects. Our own research concerns the influences of such aspects of stimulus complexity as the number of point-lights, the number of hierarchically nested jointed relations, the number of different depth planes in which motions occur, angular velocity, and the special status of sinusoidal motions.

9.7 Conclusion

If one wants to design a machine that can perform some operation, and there already exists a machine capable of this performance, then it makes good sense to try to figure out how the existing machine functions. This is, of course, the current state of affairs in the visual sciences. Computer vision research is often aimed at designing computational models that can perform perceptual feats similar to those known to be within the competencies of the human perceptual system. However, it is often the case that we know very little about *how* the human perceptual system functions, although we have good psychophysical evidence on *what* it is doing. (See Marr (1982) for a discussion of the distinction between process representations, the *what*, and algorithms, the *how*.)

We have examined the convergence in computational and psychological approaches to the problem of extracting connectivity from two-dimensional moving point-light projections of jointed objects. There exist both similarities and differences in the performance of computational models and human observers. We do not yet know whether the similarities are generalizable, or artifactually produced by the common set of problems and stimulus displays investigated in both areas. Though the differences are surely real, in view of recent work in the field, it seems reasonable to anticipate that as we learn more about the nature of human visual processing, this knowledge will assist computer vision efforts to develop more multifarious processors.

Acknowledgements

This research was supported by NICHD grant HD-16195 and Virginia CIT grant INF-85-014. Reprint requests should be addressed to Dennis Proffitt, Department of Psychology, Gilmer Hall, University of Virginia, Charlottesville, Virginia 22903-2477.

References

Andersen, G.J., and Braunstein, M.L., (1983) 'Dynamic occlusion in the perception of rotation in depth,' *Perception and Psychophysics*, vol. 34, pp. 356-362.

Attneave, F., (1971) 'Multistability in perception,' *Scientific American*, vol. 255, pp. 62-71.

Bassili, J.N., (1978) 'Facial motion in the perception of faces and of emotional expression,' *J. of Experimental Psychology: Human Perception and Performance*, vol. 4, pp. 373-379.

Beck, J., Hope, B., and Rosenfeld, A. (eds.), (1983) **Human and Machine Vision.** Academic Press, New York.

Bertenthal, B.I., and Kramer, S., (1984) 'The TMS 9918A VDP: A new device for generating moving displays on a microcomputer,' *Behavior Research Methods, Instruments and Computers*, vol. 16, pp. 388-394.

Bertenthal, B.I., and Proffitt, D.R., (1986) 'The extraction of structure from motion: Implementation of basic processing constraints,' paper presented at *The International Conference on Infant Studies, Los Angeles,* CA.

Bertenthal, B.I., Proffitt, D.R., Spetner, N.B., and Thomas, M.A., (1985) 'The development of infant sensitivity to biomechanical motions,' *Child Development*, vol. 56, pp. 531-543.

Braunstein, M.L., (1962) 'Depth perception in rotating dot patterns: Effects of numerosity and perspective,' *J. of Experimental Psychology*, vol. 64, pp. 415-420.

Braunstein, M.L., (1983) 'Perception of rotation in depth: The psychophysical evidence,' *ACM Interdisciplinary Workshop on Motion: Representation and Perception*, Toronto, Canada, pp. 119-124.

Braunstein, M.L., Andersen, G.J., and Riefer, D.M., (1982) 'The use of occlusion to resolve ambiguity in parallel projections,' *Perception and Psychophysics*, vol. 31, pp. 261-267.

Cutting, J.E., (1978) 'A program to generate synthetic walkers as dynamic point-light displays,' *Behavior Research Methods and Instrumentation*, vol. 10, pp. 91-94.

Cutting, J.E., (1983a) 'Four assumptions about invariance in perception,' *J. of Experimental Psychology: Human Perception and Performance*, vol. 9, pp. 310-317.

Cutting, J.E., (1983b) 'Perceiving and recovering structure from events,' *ACM Interdisciplinary Workshop on Motion: Representation and Perception*, Toronto, Canada, pp. 141-146.

Cutting, J.E., and Kozlowski, L.T., (1977) 'Recognizing friends by their walk: Gait perception without familiarity cues,' *Bull. of Psychonomic Society*, vol. 9, pp. 353-356.

Cutting, J.E., and Proffitt, D.R., (1981) 'Gait perception as an example of how we may perceive events,' in R. Walk and H.L. Pick, Jr. (eds.), **Intersensory Perception and Sensory Integration**, Plenum, New York.

Cutting, J.E., and Proffitt, D.R., (1982) 'The minimum principal and the perception of absolute, common and relative motions,' *Cognitive Psychology*, vol. 14, pp. 211-246.

Cutting, J.E., Proffitt, D.R., and Kozlowski, L.T., (1978) 'A biomechanical invariant for gait perception,' *J. of Experimental Psychology: Human*

Perception and Performance, vol. 4, pp. 375-372.

Doner, J., Lappin, J.S., and Perfetto, G., (1984) 'Detection of three-dimensional structure in moving optical patterns,' *J. of Experimental Psychology: Human Perception and Performance*, vol. 10, pp. 1-11.

Flinchbaugh, B.E., and Chandrasekaran, B., (1981) 'A theory of spatio-temporal aggregation for vision,' *Artificial Intelligence*, vol. 17, pp. 378-407.

Gibson, J.J., Kaplan, G.A., Reynolds, H.N., and Wheeler, K., (1969) 'The change from visible to invisible: A study of optical transition,' *Perception and Psychophysics*, vol. 5, pp. 113-116

Gogel, W.C., (1974) 'The adjacency principle in visual perception,' *Quart. J. of Experimental Psychology*, vol. 26, pp. 425-437.

Green, B.F., Jr., (1961) 'Figural coherences in the kinetic depth effect,' *J. of Experimental Psychology*, vol. 62, pp. 272-282.

Helmholtz, H., (1962) **Treatise on Physiological Optics**, Dover, New York. (Originally published in German in 1857.)

Hoffman, D.D., and Flinchbaugh, B.E., (1982) 'The interpretation of biological motion,' *Biological Cybernetics*, vol. 42, pp. 195-204.

Johansson, G., (1950) **Configuration in Event Perception**, Almqvist and Wiksell, Uppsala, Sweden.

Johansson, G., (1973) 'Visual perception of biological motion and a model for its analysis,' *Perception and Psychophysics*, vol. 14, pp. 201-211.

Johansson, G., (1976) 'Spatio-temporal differentiation and integration in visual motion perception,' *Psychological Research*, vol. 38, pp. 379-

396.

Johansson, G., (1983) 'About visual perception of biological motion, II, The movements of trees, bushes and herbs,' paper presented at the *Second International Conference on Event Perception*, Nashville, Tennessee.

Johnston, I.R., White, G.R., and Cumming, R.W., (1973) 'The role of optical expansion patterns in locomotor control,' *American J. of Psychology*, vol. 86, pp. 311 324.

Julesz, B., (1960) 'Binocular depth perception of computer generated patterns,' *Bell System Technology J.*, vol. 39, pp. 1125-1162.

Llewellyn, K.R., (1971) 'Visual guidance of locomotion,' *J. of Experimental Psychology*, vol. 91, pp. 245-261.

Marr, D., (1982) **Vision**, Freeman, San Francisco.

Marr, D., and Poggio, T., (1979) 'A computational theory of human vision,' *Proceedings of the Royal Society of London*, vol. B 204, pp. 301-328.

Martin, W.N., and Aggarwal, J.K., (1981) 'Analyzing dynamic scenes containing multiple moving objects,' in T.S. Huang (ed.), **Image Sequence Analysis**, Springer-Verlag, Berlin.

Mass, J.B., Johansson, G., Janson, G., and Runesson, S., (1971) Motion **perception I and II**, Houghton Mifflin, Boston, (film).

O'Rourke, J., and Badler, N.I., (1980) 'Model-based image analysis of human motion using constraint propagation,' *IEEE Trans. on Pattern Analysis and Machine Intelligence*, vol. PAMI-2, pp. 524-536.

Perkins, D.N., (1982) 'The perceiver as organizer and geometer,' in J. Beck (ed.), **Organization and Representation in Perception**, Erlbaum, Hillsdale, NJ.

Proffitt, D.R., (1981) 'Kinetic and proximity influences in perceiving structure in revolving point-lights,' paper presented at *22nd Annual Meeting of the Psychonomic Society*, Philadelphia.

Proffitt, D.R., (1983) 'Factors affecting multistability in moving point-light displays,' Paper presented at *24th Annual Meeting of the Psychonomic Society*, San Diego.

Proffitt, D.R., and Bertenthal, B.I., (1984) 'Perceiving 3-D patterns of connectivity on moving point-light projections,' Paper presented at *25th Annual Meeting of the Psychonomic Society*, San Antonio, TX.

Proffitt, D.R., Bertenthal, B.I., and Roberts, R.J., Jr., (1984) 'The role of occlusion in reducing multistability in moving point-light displays,' *Perception and Psychophysics*, vol. 36, pp. 215-232.

Proffitt, D.R., and Cutting, J.E., (1979) 'Perceiving the centroid of configurations on a rolling wheel,' *Perception and Psychophysics*, vol. 26, pp. 389-398.

Proffitt, D.R., and Cutting, J.E., (1980a) 'An invariant for wheel-generated motions and the logic of its determination,' *Perception*, vol. 9, pp. 435-449.

Proffitt, D.R., and Cutting, J.E., (1980b) 'Perceiving the centroid of curvilinearly bounded rolling shapes,' *Perception and Psychophysics*, vol. 28, pp. 484-487.

Proffitt, D.R., Thomas, M.A., and O'Brien, R.G., (1983) 'The role of contour and luminance distribution in determining perceived centers within shapes,' *Perception and Psychophysics*, vol. 33, pp. 63-71.

Regan, D., and Beverley, K.I., (1982) 'How do we avoid confounding the direction we are looking and the direction we are moving?' *Science*, vol. 215, pp. 194-196.

Restle, F., (1979) 'Coding theory of the perception of motion configurations,' *Psychological Review*, vol. 86, pp. 1-24.

Roach, J.W., and Aggarwal, J.K., (1980), 'Determining the movement of objects from a sequence of images,' *IEEE Trans. on Pattern Analysis and Machine Intelligence*, vol PAMI-2, no. 6, pp. 554-562.

Spetner, N.B., and Bertenthal, B.I., (1983) 'The development of sensitivity to facial expressions of emotion in point-light displays,' Paper presented at *the Virginia Forum for Developmental Research*, Richmond, Virginia.

Ullman, S., (1979) **The interpretation of Visual Motion**, MIT Press, Cambridge, MA.

Ullman, S., (1985) 'Maximizing rigidity: The incremental recovery of 3-D structure from rigid and nonrigid motion,' *Perception*, vol. 13, pp. 255-274.

Wallach, H., and O'Connell, D.N., (1953) 'The kinetic depth effect,' *J. of Experimantal Psychology*, vol. 45, pp. 204-217. (Also in H. Wallach, **On Perception**, Quadrangle, New York, 1976.)

Webb, J.A, and Aggarwal, J.K., (1982) 'Structure from motion of rigid and jointed objects,' *Artificial Intelligence*, vol. 19, pp. 107-130.

Wertheimer, M., (1938) 'Laws of organization in perceptual forms,' in **A Source Book of Gestalt Psychology**, W.D. Ellis (ed.), Routledge and

Kegan Paul, London, (Originally published in German in 1923.)

CHAPTER 10

Algorithms for Motion Estimation Based on Three-Dimensional Correspondences

S. D. Blostein

T. S. Huang

10.1 Introduction

The goal in the motion estimation problem is to determine the transformation between two three-dimensional positions of a rigid body that is undergoing some arbitrary motion. Throughout this discussion it will be assumed that it is possible to acquire three-dimensional positional information of points located on the rigid body at two separate time instances. This may be accomplished through the use of a stereo camera setup (Barnard and Fischler, 1982). Alternatively, three-dimensional positional data can be obtained explicitly from a laser range finder. In general, points selected from the rigid body will tend to correspond to special geometrical features, such as corners or identifiable surface markings, and must often be extracted by special low level processing tasks (Gu and Huang, 1984). Such feature points will be invariant to the particular position sensing techniques used, and thus may be reliably located independent of object position and orientation.

The determination of motion parameters provides all the necessary information about the displacement of the object in three-dimensional space and can be viewed as a two-step process. First, the correspondence between some extracted feature points common to both views of the object must be established, i.e., a set of points whose position is known at each of the two time instances must be found. Second, this set of point correspondences must be input to a *motion estimation algorithm* that determines enough parameters to uniquely define a change in three-dimensional position. Throughout this chapter the concern will be focussed on the latter issue: it will be assumed that point correspondences have been determined and that the goal is to extract motion parameters from such data. It should be noted that the results from the second step can be used to improve the solution to the first. Assuming that the object is a rigid body, the displacement transformation should hold true for all points in the data set, and thus tentative point matches may be verified once the motion parameters are determined.

In order to formulate this problem mathematically there must be a definition of a general displacement that allows for all possible transformations that a rigid body may undergo. In addition, the minimum number of point locations that uniquely determine a rigid body's position must be determined. This will serve as a lower bound on the number of point correspondences needed to determine the motion parameters uniquely. This number of point correspondences is given in the following theorem.

THEOREM 1: Three noncollinear points on a rigid body are necessary and sufficient to uniquely determine its position and orientation in three-dimensional space.

PROOF: Suppose that the rigid body fixed at position, Q, contains two points, A and B. If the only known points to lie on the rigid body were A and B, or one of these, it is obvious that the rigid body may be rotated about an axis through A and B by some arbitrary angle, $0 \leq \phi < 2\pi$, to position, Q'. In general, Q and Q' contain points that have been displaced and hence specify different three-dimensional positions. Thus the position of the rigid body cannot be determined uniquely from just A and B or one of these. Now if a third noncollinear point, C, were also known on the body, there would exist a fixed axis through A and B as well as a fixed point, C. Clearly, it is no longer possible to rotate the rigid body by some nonzero angle without altering the location of C. Thus, three points are both necessary and sufficient to fix a rigid body in three-dimensional space.

■

For a more extensive presentation, see (Ball, 1876).

We next define a general displacement by the equation,

$$p' = Rp + t, \tag{1}$$

where p and p' represent a point on the rigid body at time instances, T_1 and T_2, respectively. R is a 3×3 matrix specifying a rotation of some angle about an axis arbitrarily oriented in three-dimensional space and passing through the origin of the coordinate system. This rotation matrix is a member of the special orthogonal group, SO(3), having the properties: $R R^t = I$, the identity matrix, and Det(R), the determinant of R, equal to 1. Finally, t is a translation vector specifying an arbitrary shift of the point, p, after rotation. In component form, then, Eq. 1 is

$$\begin{bmatrix} p'_x \\ p'_y \\ p'_z \end{bmatrix} = \begin{bmatrix} r_{11} & r_{12} & r_{13} \\ r_{21} & r_{22} & r_{23} \\ r_{31} & r_{32} & r_{33} \end{bmatrix} \begin{bmatrix} p_x \\ p_y \\ p_z \end{bmatrix} + \begin{bmatrix} t_x \\ t_y \\ t_z \end{bmatrix}. \tag{2}$$

10.2 Direct Linear Method

From the above, one may initially observe that there are twelve free parameters for which to solve, namely R (nine parameters) and t (three parameters). Accordingly one can express Eqs. 1 and 2 by three, four-dimensional systems of linear equations as follows:

$$
\begin{bmatrix}
p_{1x} & p_{1y} & p_{1z} & 1 \\
p_{2x} & p_{2y} & p_{2z} & 1 \\
p_{3x} & p_{3y} & p_{3z} & 1 \\
p_{4x} & p_{4y} & p_{4z} & 1
\end{bmatrix}
\begin{bmatrix}
r_{11} \\
r_{12} \\
r_{13} \\
t_x
\end{bmatrix}
=
\begin{bmatrix}
p'_{1x} \\
p'_{2x} \\
p'_{3x} \\
p'_{4x}
\end{bmatrix}
\tag{3}
$$

$$
\begin{bmatrix}
p_{1x} & p_{1y} & p_{1z} & 1 \\
p_{2x} & p_{2y} & p_{2z} & 1 \\
p_{3x} & p_{3y} & p_{3z} & 1 \\
p_{4x} & p_{4y} & p_{4z} & 1
\end{bmatrix}
\begin{bmatrix}
r_{21} \\
r_{22} \\
r_{23} \\
t_y
\end{bmatrix}
=
\begin{bmatrix}
p'_{1y} \\
p'_{2y} \\
p'_{3y} \\
p'_{4y}
\end{bmatrix}
\tag{4}
$$

$$
\begin{bmatrix}
p_{1x} & p_{1y} & p_{1z} & 1 \\
p_{2x} & p_{2y} & p_{2z} & 1 \\
p_{3x} & p_{3y} & p_{3z} & 1 \\
p_{4x} & p_{4y} & p_{4z} & 1
\end{bmatrix}
\begin{bmatrix}
r_{31} \\
r_{32} \\
r_{33} \\
t_z
\end{bmatrix}
=
\begin{bmatrix}
p'_{1z} \\
p'_{2z} \\
p'_{3z} \\
p'_{4z}
\end{bmatrix}.
\tag{5}
$$

These expressions imply that four correspondences are needed to solve the above: (p_1, p'_1), (p_2, p'_2), (p_3, p'_3), and (p_4, p'_4), where $(p_i, p'_i) \equiv ((p_{ix}, p_{iy}, p_{iz}), (p'_{ix}, p'_{iy}, p'_{iz}))$. The condition of the four-dimensional square matrices, in Eqs. 3, 4 and 5 is dependent entirely on the positions of the points chosen on the rigid body. For the matrix in Eq. 1 to have full rank, four point correspondences which do not all lie on the same plane are needed. Note that the numerical stability is independent of the choice of motion, a desirable property.

10.3 Method Based on Translation Invariants

We can associate a unit vector, $\hat{m}$, with any three-dimensional line. We define $\hat{m}$,

$$\hat{m} = \frac{p_2 - p_1}{|\ p_2 - p_1\ |}, \tag{6}$$

to be the relative orientation of points, p_1 and p_2, on a rigid body. Now if the body undergoes a pure translation, these $\hat{m}$ parameters do not change. Only when the body rotates are the $\hat{m}$-vectors transformed. If two points on the body, p_i and p_{i+1}, which undergo the same displacement due to rigidity of the body, move to respective positions, p'_i and p'_{i+1}, then

$$p'_i = Rp_i + t$$

and

$$p'_{i+1} = Rp_{i+1} + t.$$

Subtraction eliminates the translation, t, and using the rigidity constraint yields

$$\frac{p'_{i+1} - p'_i}{|\ p'_{i+1} - p'_i\ |} = R\ \frac{p_{i+1} - p_i}{|\ p_{i+1} - p_i\ |},$$

which from Eq. 6 is

$$\hat{m}'_i = R\ \hat{m}_i \tag{7}$$

for point correspondences, $1 \le i \le 3$, where $\hat{m}'_i$ is defined analogously to

Eq. 6.

We can perform the same rearrangement as before and solve three, 3 × 3 systems to get R, and afterward obtain t by substitution into Eq. 1. Note that the solution of the systems of linear equations are now of lower dimension than in the systems of Eqs. 3, 4 and 5. Finally, in order for a unique solution to exist, the 3 × 3 matrix of unit m̂-vectors must be of full rank, i.e., the three m̂-vectors must not be coplanar. As a result, four point correspondences are needed, as in the *direct linear method* of Section 10.2, and any motion will yield a unique solution.

It must be pointed out that a general three-dimensional displacement can be described in far fewer than 12 parameters. As stated earlier the rotation matrix R is an orthogonal matrix which can be expressed in terms of a rotation axis, $\hat{n} = [\, n_x, n_y, n_y \,]^t$, and a rotation angle, ϕ, about $\hat{n}$. The following derivation will obtain the structure of R using a geometrical setup similar to MacMillan (1936).

In Figure 1, point, P, is rotated by angle, ϕ, to P′, with p and p′ the respective position vectors relative to origin, O. Through O passes the axis of rotation, $\hat{n}$. At the center of the circle shown in Figure 1 lies point, A, such that OAP and OAP′ are right angles. Let segment, AB, be directed 90° ahead of segment, AP, in the same plane as A, P, and P′. Then AB = $\hat{n} \times p$ and AP = $-\hat{n} \times (\hat{n} \times p)$, since AB and AP both have the same length by construction. From these relations the position of point, P′, can be expressed by

$$
\begin{aligned}
p' &= OA + AP' \\
&= p - AP + AP\cos\phi + AB\sin\phi \\
&= p + \hat{n} \times (\hat{n} \times p) - \hat{n} \times (\hat{n} \times p)\cos\phi + (\hat{n} \times p)\sin\phi.
\end{aligned}
$$

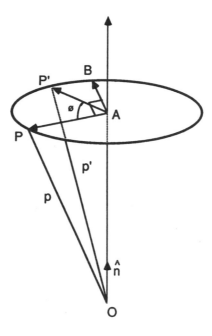

Figure 1. A three dimensional rotation of point p to p′ about axis n̂.

Rearranging and using the fact that $\hat{n} \times (\hat{n} \times p) = (\hat{n} \cdot p) \hat{n} - p$ gives the result,

$$p' = (\hat{n} \times p)\sin\phi + p \cos\phi + (1 - \cos\phi) \hat{n} (\hat{n} \cdot p). \qquad (8)$$

Replacing the cross product term by a product with skew-symmetric matrix N and using the distributive rule for the dot product term yields the matrix equation,

$$p' = \left[\sin\phi \, N + \cos\phi \, I + (1 - \cos\phi) \, \hat{n} \, \hat{n}^T \right] p, \qquad (9)$$

$$\text{where} \quad N = \begin{bmatrix} 0 & -n_z & n_y \\ n_z & 0 & -n_x \\ -n_y & n_x & 0 \end{bmatrix}.$$

The bracketed term multiplying p in Eq. 9 is the rotation matrix R, which expands into component form as

$$R = \begin{bmatrix} (n_x^2 - 1)c + 1 & n_x n_z c - n_z s & n_x n_z c + n_y s \\ n_y n_x c + n_y s & (n_y^2 - 1)c + 1 & n_y n_z c - n_x s \\ n_z n_x c - n_y s & n_z n_y c + n_x s & (n_z^2 - 1)c + 1 \end{bmatrix} \qquad (10)$$

where $c = (1 - \cos\phi)$ and $s = \sin\phi$. Since $\hat{n}$ is a unit vector, R can be expressed in terms of three parameters which, in addition to the translation vector, yield six parameters that describe a general displacement in three-dimensional space. Alternatively, the R matrix and t vector can be determined by solving for these 6 parameters rather than 12 as in the previous case. The price paid is that the displacement is not linear in $\hat{n}$, ϕ, and t, and solution is less straightforward.

On inspection of Eq. 10, it may seem quite forbidding to solve for the rotation axis and angle to get R. However, two successful nonlinear approaches have made it possible to express the rotation axis, $\hat{n}$, and the rotation angle, ϕ, directly in closed form in terms of the point correspondences. These closed form solutions have computational advantages over the linear methods described previously, since no equations need be solved.

10.4 Axis-Angle Method

The intermediate results from the previous discussion can be used to obtain the rotation axis, $\hat{n}$, from two noncollinear $\hat{m}$-vectors, $\hat{m}_1$ and $\hat{m}_2$. Since the rotation takes place about axis, $\hat{n}$, the angle between any $\hat{m}$-vector and $\hat{n}$ should remain constant during the transformation. That is,

$\hat{n} \cdot \hat{m} = \hat{n} \cdot \hat{m}'$ or $\hat{n} \cdot (\hat{m} - \hat{m}') = 0$. In other words, $\hat{n}$ must be orthogonal to $(\hat{m} - \hat{m}')$ for any $\hat{m}$-vector correspondence. Taking any two of these $\hat{m}$-vector correspondences, the axis direction, n_{dir} can then be found by

$$n_{dir} = (\hat{m}_1' - \hat{m}_1) \times (\hat{m}_2' - \hat{m}_2), \tag{11}$$

with $\hat{n} = n_{dir} / \mid n_{dir} \mid$.

Once the rotation axis has been determined, the rotation angle can be found. Since $\hat{m}$ and $\hat{m}'$ are related by a pure rotation, Eq. 7, they are also related by Eq. 8 as

$$\hat{m}' = (\hat{n} \times \hat{m}) \sin\phi + \hat{m} \cos\phi + (1 - \cos\phi) \hat{n} (\hat{m} \cdot \hat{m}).$$

Taking the dot product of both sides with $\hat{m}$ to eliminate the sine term and then solving for $\cos\phi$ gives

$$\cos\phi = 1 - \frac{1 - \hat{m}' \cdot \hat{m}}{1 - (\hat{n} \cdot \hat{m})^2}. \tag{12}$$

Similarly, taking the dot product of both sides with $\hat{m}'$ to eliminate the cosine term and rearranging yields

$$\sin\phi = \frac{(\hat{m} \times \hat{n}) \cdot \hat{m}'}{1 - (\hat{n} \cdot \hat{m})^2}. \tag{13}$$

From $\cos\phi$, $\sin\phi$, and $\hat{n}$, the rotation matrix, R, can be computed by substitution into Eq. 10. Finally, the translation vector, t, can be solved for by substitution since R is known, i.e., $t = p' - Rp$. The entire procedure requires only additions and multiplications and obviously much fewer manipulations than do the solutions of systems of linear equations, such as required in the previous methods.

For this case it is no longer true that the numerical stability is purely a function of the locations of the point correspondences. To satisfy Theorem 1, the points must not be collinear in order to uniquely specify the position of the rigid body, implying that three point correspondences are needed. However, by observing Eq. 11, as $\hat{m}_1 \to \hat{m}'_1$ and $\hat{m}_1 \to \hat{m}'_1$, i.e., the pure translation case, $n_{dir} \to 0$, causing $\hat{n}$ to become poorly defined. Moreover, the error in computing Eq. 11 propagates into the angle computation of Eqs. 12 and 13, and becomes unbounded. Thus, as the motion degenerates to a pure translation, this method's accuracy decreases significantly. A partial remedy to this problem can be found by taking special care when the angels are small (Paul, 1981), but the same basic problem arises nevertheless.

10.5 The Screw Decomposition Method

As an alternative approach, a general three-dimensional displacement can be expressed in terms of a different set of parameters than the six used for $\hat{n}$, ϕ, t_x, t_y, and t_z. [Since $\hat{n}$ has unit magnitude n_z is determined by n_x and n_y.] Instead, a motion can be parameterized by six *screw displacement* parameters, $\hat{s}$, s_0, θ, and d.

Note that in the previous case the rotation angle, ϕ, was about an axis, $\hat{n}$, through the origin of the coordinate system. This representation is often unnatural in the real world since many rigid bodies appear to rotate about an axis *not* through the origin of the frame of reference, but rather seem to undergo a *screw motion* about an axis parallel to the direction of translation. In other words, the rotation axis need not be fixed to pass through the origin but can instead be displaced a distance from the origin, even to pass through the object itself. Moreover, the translation can be defined as occurring purely in the direction of this screw axis. The fact that a general three-dimensional displacement can be represented as a *unique* screw displacement is a statement of Chasles' Theorem (Bottema and Roth, 1979). It turns out that this screw decomposition is characterized by six parameters, $s_x, s_y, s_{0x}, s_{0y}, \theta$, and d, where $\hat{s} \equiv [s_x, s_y, s_z]^t$ is a unit vector representing the orientation of the screw axis, $s_0 \equiv [s_{0x}, s_{0y}, s_{0z}]^t$ is the perpendicular distance of the screw axis to the origin, θ is the screw rotation angle about $\hat{s}$ located at s_0, and d is the translation along $\hat{s}$. Note that s_0 and $\hat{s}$ are perpendicular and $\hat{s}$ is a unit vector, so $s_0 \cdot \hat{s} = 0$ and $\hat{s} \cdot \hat{s} = 1$. Thus, only four parameters are needed to describe $\hat{s}$ and s_0. The inclusion of d and θ brings the total number of *screw parameters* to six.

First, the relationship between a screw displacement and the general displacement given in Eq. 1 must be determined. This will result in a method that solves first for the screw parameters, and then expresses R and t in terms of the computed parameters. Before presenting this solution method, the relationship between the two displacements will be shown. Recall that a screw displacement is a rotation of 0 about an axis, $\hat{s}$, located at point, s_0, followed by a translation, d, along that axis. The displacement from p to p' can be expressed in the form of Eq. 1 if we define the shifted points,

$$q = p - s_0 \tag{14a}$$

and

$$q' = p' - s_0, \tag{14b}$$

resulting in a displacement from q to q' about an axis through a new origin, $O' \equiv [-s_{0x}, -s_{0y}, -s_{0z}]$, followed by the screw translation. We have

$$q' = R_s q + d\hat{s} \tag{15a}$$

or

$$(p' - s_0) = R_s (p - s_0) + d\hat{s}, \tag{15b}$$

so

$$p' = R_s p + (I - R_s) s_0 + d\hat{s}. \tag{15c}$$

Equating the above terms with those in Eq. 1 gives

$$R = R_s \tag{16a}$$

and

$$t = (I - R_s) s_0 + d\hat{s} \tag{16b}$$

From Eqs. 16a and 16b note that the screw rotation matrix R_s is identical to the R matrix defined originally, thus $\phi = \theta$ and $\hat{n} = \hat{s}$. The translation vector, t, is composed of two components, one corresponding to a translation along the screw axis and another corresponding to the change in location of the screw axis relative to the origin of the original system.

A solution method for finding $\hat{s}$, s_0, and θ has been proposed (Bottema and Roth, 1979) based on Rodrigues' formula for a general screw displacement:

$$p' - p = \left[\tan\left[\frac{\theta}{2}\right]\right] \hat{s} \times (p' + p - 2s_0) + d\hat{s}. \tag{17}$$

The above equation can be used to determine the screw parameters given point correspondences, (p_1, p_1'), (p_2, p_2'), and (p_3, p_3'). Writing Eq. 17 once for the first point and once for the second point and subtracting the two equations yields

$$(p_1' - p_2') - (p_1 - p_2) = \left[\tan\left[\frac{\theta}{2}\right]\right] \hat{s} \times \left[(p_1' - p_2') + (p_1 - p_2)\right].$$

Normalizing both sides, the above can be expressed in terms of $\hat{m}$-vectors:

$$\hat{m}_1' - \hat{m}_1 = \left[\tan\left(\frac{\theta}{2}\right)\right]\hat{s} \times (\hat{m}_1' + \hat{m}_1).$$

Taking the cross product of both sides with $[(p_{3'} - p_{2'}) - (p_3 - p_2)]$ and noting from the previous discussion that a vector in direction $(\hat{m}_2' - \hat{m}_2)$ is perpendicular to the rotation axis, the screw axis and angle can be solved for by

$$\tan\left[\frac{\theta}{2}\right]\hat{s} = \frac{[\hat{m}_2' - \hat{m}_2] \times [\hat{m}_1' - \hat{m}_1]}{[\hat{m}_2' - \hat{m}_2] \cdot [\hat{m}_1' + \hat{m}_1]} \tag{18}$$

and normalizing the left-hand side. Also, $\cos\theta$ and $\sin\theta$ can be determined from $\tan\left[\dfrac{\theta}{2}\right]$, i.e., the magnitude of the right-hand side, by the trigonometric identities,

$$\cos\theta = \frac{1 - \tan^2\dfrac{\theta}{2}}{1 + \tan^2\dfrac{\theta}{2}}$$

and

$$\sin\theta = \frac{2\tan\dfrac{\theta}{2}}{1 + \tan^2\dfrac{\theta}{2}}.$$

The location of the screw axis, s_0, can be obtained using one of the point correspondences by taking the cross product of $\hat{s}$ with Rodrigues' formula, Eq. 17, and noting that $\hat{s} \cdot s_0 = 0$:

$$s_0 = \frac{1}{2} \left[p_1 + p_1' + \frac{(\S \times (p_1' - p_1))}{\tan\frac{\theta}{2}} - \S \cdot (p_1 + p_{1'}) \, \S \right].$$ (19)

Finally, taking the dot product of $\S$ with each side of Eq. 17 and simplifying yields

$$d = \S \cdot (p_1' - p_1),$$ (20)

where d is the displacement projected along the screw axis. R and t can be found by substituting quantities from Eqs. 18, 19 and 20 into Eqs. 16a and 16b. In short, the motion parameters have been recovered from three non-collinear points. Note that in Eq. 18, $\S$ becomes unspecified as θ approaches zero, i.e., the pure translation case, and becomes unbounded as θ approaches 180°. Also, the location, s_0, of $\S$ in Eq. 19 becomes unbounded as θ approaches zero.

Up to this point several methods have been presented that solve for the rotation and translation parameters. Two of these are linear methods the *direct linear method*, Eqs. 3, 4 and 5 which solved 4 × 4 systems of point correspondences; and the *translation invariant method*, a rearrangement of Eq. 7, which first involved solving 3 × 3 systems of $\dot{m}$-parameters for R and then finding t by substitution. The other two methods, the *axis-angle* method and the *screw decomposition* method, were nonlinear and did not involve solving systems of equations. Before comparing the numerical differences between these methods, it is pertinent to state the conditions under which each of the methods fails. There are two criteria which affect the numerical properties of the solutions. The first consideration is the choice of point correspondences obtained from the matching process. The second is the behavior of each method under different possible motions. Obviously, a unique solution for any of the methods requires that the set of points chosen not be collinear.

10.6 Improved Motion Estimation Algorithms

In the preceding discussion, methods have been described which solve for the rotation matrix and the translation vector given a set of point correspondences found from two different time instances. The linear methods, although somewhat more expensive computationally, have the decisive advantage of numerical accuracy independent of motion. The drawback seems to be insuring that the matrix resulting from the point correspondences selected will be of full rank. Also, the linear methods introduced so far use four points instead of the minimum of three required for uniqueness as specified in Theorem 1. To overcome the problem of supplying the linear methods with this extra point correspondence a "pseudo-correspondence" can be artificially constructed in the case of the *direct linear method*. For the *translation invariant method* the same procedure can be adopted by finding a third $\hat{m}$-vector orthogonal to the other two. One would expect a better improvement in the latter method for two reasons. First, the $\hat{m}$-vectors are normalized, and secondly, the system is of lower dimension. This has been shown experimentally in the simulation results presented in Section 10.8. Both of these above ingredients help reduce a matrix's nearness to singularity, i.e., lower its condition number. In particular, the construction of an "orientation robust basis" can proceed by first recalling the relationship between $\hat{m}$-vector correspondences, $\hat{m}'_k = R\hat{m}_k$, for $1 \leq k \leq 3$. The first row of the R matrix can be determined in component form by

$$
\begin{bmatrix}
m_{1x} & m_{1y} & m_{1z} & 1 \\
m_{2x} & m_{2y} & m_{2z} & 1 \\
m_{3x} & m_{3y} & m_{3z} & 1
\end{bmatrix}
\begin{bmatrix}
r_{11} \\
r_{12} \\
r_{13}
\end{bmatrix}
=
\begin{bmatrix}
m'_{1x} \\
m'_{2x} \\
m'_{3x}
\end{bmatrix},
\qquad (21)
$$

while the other two rows of R can be determined by replacing the right-hand side by the y and z components of the primed $\hat{m}$-vectors. A very low condition number can be achieved by setting $\hat{m}_3 = \hat{m}_1 \times \hat{m}_2$ and

$\hat{m}'_3 = \hat{m}'_1 \times \hat{m}'_2$. These "artificial" vectors $\{\hat{m}_3, \hat{m}'_3\}$ are generated to complete the bases, $\{\hat{m}_1, \hat{m}_2, \hat{m}_3\}$ and $\{\hat{m}'_1, \hat{m}'_2, \hat{m}'_3\}$, and thus, to span the three dimensions. This method will be called, the *improved translation invariant method.*

10.7 Comparing the Linear and Nonlinear Methods

Four distinct methods for solving for the rotation and translation parameters, each requiring three, three-dimensional point correspondences, have been established: (1) the improved direct linear method, (2) the improved translation invariant method, (3) the angle-axis method, and (4) the screw decomposition method. The first two methods involve solving systems of linear equations and are thus termed, linear methods, while the second two methods simply require explicit evaluation of nonlinear expressions to first obtain the rotation axis and angle. Suppose that the criterion for evaluating the performance of these four methods was to minimize relative error in the motion parameters. For a linear system of equations, Ax = b, the condition number of matrix A, cond(A), is defined as the ratio of the maximum eigenvalue of A to the minimum eigenvalue of A. Alternatively, the condition number can be expressed as

$$\text{cond(A)} \equiv \frac{\max_{x} \frac{|Ax|}{|x|}}{\min_{x} \frac{|Ax|}{|x|}},$$

where either the L1 or L2 norm may be used in the above expression. It can be shown (Forsythe and Malcolm, 1977) that

$$\frac{|\Delta x|}{|x|} \leq \text{cond(A)} \frac{|\Delta b|}{|b|}. \tag{22}$$

Note that the condition number is simply the magnification factor of the relative error in the right-hand side relative to the error in the solution vector. In the linear methods discussed, however, there is not only a perturbation possible in the right-hand side, but also in the matrix itself. In Eq. 21 both the matrix and the right-hand side contain noise. This is known as an *approximate* system of linear algebraic equations (Kuperman, 1971). In matrix form, this system can be expressed as

$$MR = M',\tag{23}$$

where M stands for the matrix of observed $\hat{m}$-vectors embedded in noise at time T1, R is the rotation matrix, and M' is an observed vector of $\hat{m}$-components at T2. By decomposing each observed quantity into the sum of its true value plus a noise term, the approximate system of equations can be transformed into a linear system with a right-hand side perturbation,

$$(M + \Delta M)\, R = (M' + \Delta M'),$$

or equivalently,

$$M\,R = M' + (\Delta M' - \Delta M\, R).\tag{24}$$

The condition number, then, can be effectively used as a performance measure for the linear motion estimation methods considered in this discussion.

Performance of the nonlinear methods cannot be measured as easily as in the case of the linear methods due to their rather intractable behavior. As an extension to the discussion of the accuracy of the nonlinear methods at the end of Section 10.5, it is easy to show that sensitivity to observation noise increases rapidly as the angle of rotation becomes small. In the *axis-angle method*, for example, the rotation axis direction is first obtained from Eq. 11, so that as the motion approaches a pure translation, $\hat{m}_i \rightarrow \hat{m}'_i$, i = 1, 2. The observed value of the rotation axis contains a very small true parameter value compared to the noise component, and thus the rotation axis becomes undefined as the motion approaches a pure translation causing the rotation matrix to consist purely of noise. The linear methods, however, do not exhibit this problem since they solve for the R matrix first which tends to a unit matrix as the motion becomes a pure translation.

Finally, the orthogonality principle in linear estimation method can be used. It is well known that a least-mean-square error criterion can be satisfied if and only if the class of estimators has the property of being uncorrelated with the estimation error (Papoulis, 1965). Taking on a geometric interpretation, it is desirable that each of the two sets of position estimates (of points on a rigid body) be represented by an orthonormal basis so as to be orthogonal to random noise in three-dimensional position. Assuming a measurement sensor noise that is independent of three-dimensional orientation, the best possible estimate of orientation would be to construct an orthonormal basis from the $\hat{m}$-vectors at both T1 and T2. In fact, the *improved translation invariant method* is a very close approximation to such a construction. To obtain a complete orthonormalization of the $\hat{m}$-vectors the Gram-Schmidt technique (Brogan, 1982) can be used. In practice, complete orthonormalization only slightly out performs the *improved translation invariant method*, but requires significantly more computation.

10.8 Simulation Results for Three-Point Methods

The methods so far described have been simulated and the results for a quantized, three-dimensional space of $1024 \times 1024 \times 1024$ points are shown here. Corresponding results at $(512)^3$ and $(256)^3$ quantizations yield, respectively, about double and quadruple the average relative errors obtained for the "1024" case. For the simulation a general motion consisting of a rotation of $23°$ about an axis with direction, (0.7, 0.5, 0.51), and a translation of (63, 35, -150) in this quantized environment was used.

Average relative errors of the nine rotation matrix components and the three translation components were computed over 1000 sets of randomly generated points and their corresponding points that were transformed according to the motion specified. This experiment was repeated for several different motions, and the results obtained were similar to those described here with one notable exception: for motions that contained a very small or nonexistent amount of rotation, the nonlinear methods performed very poorly, as expected. The reasoning behind measuring *average* quantities was to evaluate the performance of a solution method independent of a particular configuration of point correspondences. Unfortunately, spatially localized visible points encountered in practical situations are constrained as to their spatial distribution, a factor not taken into account by the simulation: points were simply chosen from the universe space with uniform distribution. The motion estimations giving rise to the largest errors, i.e., the worst case results over the 1000 "trials", are also presented. These worst case results tend to arise when the three points lie in nearly a straight line causing the problem to be ill conditioned.

The results are in the two tables shown. Table 1 relates to the discussion on improving the linear methods, Section 10.6, and consists of a comparison between three variations of the direct linear method: (1) the straightforward approach first mentioned in Section 10.2 that uses *four* correspondences; (2) same method as (1), except that four of the correspondences chosen were from an over-determined case where *five* correspondences were available (the matrix of lowest condition number was used); and (3) the *improved direct linear method* from Section 10.6 that uses only *three* point correspondences. The results show that a point matrix with a lower condition number will consistently yield better performance. Four possible matrices can be used since five correspondences are available and four are used. The matrix that possessed the smallest condition number was chosen.

Table 2 compares the two improved versions of linear methods from Section 10.6 as well as the two nonlinear methods discussed in Sections 10.4 and 10.5. Note that the *improved translation invariant method* gives the best results, with Each of these methods listed in Table 2 requires three point correspondences, and the simulation for each was run on the same sets of point correspondences.

From the results obtained, these improved methods performed satisfactorily and at very low computational cost, i.e., the best method yielded average errors of less than one percent for the motion described. This is encouraging since the *improved translation invariant method* has accuracy independent of the motion to be determined. However, a true bottleneck lies in the problem in obtaining these point correspondences in the first place, which has not been discussed here. Existing procedures for finding correspondences are slow and not yet completely reliable.

10.9 Some Recent Related Results

The methods described so far are basic techniques of extracting motion parameters from position data. In many practical situations, the measurement noise in obtaining the point correspondences may be significant. In such cases the "rotation" matrix found will generally no longer represent a rotation of a rigid body and may be far from orthogonal. This is particularly distressing if it is known a priori that the object remains rigid over the time interval between corresponding positions. Thus, a technique for extracting motion parameters in the presence of noise would be to constrain the matrix R to represent a true rotation. Given N point correspondences, $\{p_i, p'_i\}$, $i = 1, 2, ..., N$, related by

$$p'_i = Rp_i + t + N_i, \tag{25}$$

where the p_i and p'_i are 3 by 1 column vectors and N_i is a noise vector, a reasonable criterion would be to find R and t to minimize

$$\Sigma^2 = \sum_{i=1}^{N} |\,|p'_i - (Rp_i + t)|\,|^2 \tag{26}$$

under the constraint that R is a rotation matrix.

Method Used	Avg % errors in rotation matrix components	Avg % errors in translation components	% error of worst component
direct linear method	3.10	22.42	273.0
same; "overdetermined"	0.464	1.79	35.6
"improved" version	0.337	1.45	21.7

Table 1.

Method Used	Avg % errors in rotation matrix components	Avg % errors in translation components	% error of worst component
improved direct linear	0.337	1.45	21.7
improved trans inv	0.279	1.11	12.7
axis-angle	4.27	9.33	64.9
screw decomposition	2.65	13.7	173.0

Table 2.

Three algorithms for finding the solution are known to exist. One of them is iterative and is described in (Huang, Blostein, and Margerum, 1986). The two others are noniterative and can be described in a closed form. The first (Faugeras and Hebert, 1983) is an elegant approach using quaternions and requires the computation of the minimum eigenvalue-eigenvector pair of a 4 by 4 matrix. The second (Arun, Huang and Blostein, 1987) uses the singular value decomposition (SVD) of a 3 by 3 matrix. While all three methods obtain the same solution to a given problem, the latter two methods have somewhat of a computational advantage over the first. For a comparison of the different computation time requirements see (Arun, Huang and Blostein, 1987).

In many cases a small subset of the N points may be extremely inaccurate. For example, the point correspondences may be derived from a "matching" algorithm where false matches or missing points can occur between the two time instances. In such situations, the need to combat the effects of "outliers" in the data arises. Some point correspondences may be quite reliable, while others should not be considered at all in the motion estimation process. A priori, of course, it is not known which points, if any, are outliers. The "least-square" algorithms of Eq. 26 are inadequate since equal weight is placed upon each of the point correspondences. Robust techniques for handling outliers have been developed using iteratively reweighted least-squares as well as random sample consensus (RANSAC). For a discussion on these, see (Blostein and Huang, 1985).

References

Arun, S., Huang, T.S., and Blostein S.D., (1987) 'Least-squares fitting of two 3-D point sets,' *IEEE Trans. on Pattern Analysis and Machine Intelligence,* vol. 9, no, 5.

Ball, R., (1876) **Theory of Screws,** Hodges, Foster and Co., Dublin.

Barnard, S., and Fischler, M.A., (1982) 'Computational stereo.,' *Computing Surveys,* vol. 14, no. 4.

Blostein, S.D., and Huang, T.S., (1985) 'Motion estimation based on stereo sequences,' Tech. Rep. T-168, Coordinated Science Lab, University of Illinois.

Bottema, O., and Roth, B., (1979) **Theoretical Kinematics,** North Holland, Amsterdam, pp. 56-62.

Brogan, W.L., (1982) **Modern Control Theory,** Prentice-Hall, Englewood Cliffs, NJ, pp. 69-70.

Forsythe, G., and Malcolm, M., (1977) **Computer Methods for Mathematical Computations,** Prentice-Hall, Englewood Cliffs, NJ, pp. 41-48.

Faugeras, O.D., and Hebert, M., (1985) 'A 3-D recognition and positioning algorithm using geometrical matching between primative surfaces,' *Proc. Int. Joint Conf. on Artificial Intelligence,* Karlshure, FR Germany, pp. 998-1002.

Gu, W.K., and Huang, T.S., (1984) 'Connected edge extraction from perspective views of a polyhedron,' Tech. Rep., University of Illinois.

Huang, T.S., Blostein, S.D., (1985) 'Robust algorithms for motion estimation based on two sequential stereo image pairs,' *IEEE Conf. on Computer Vision and Pattern Recognition,* San Francisco, CA, pp. 518-523.

Huang, T.S., Blostein S.D., and Margerum, E.A., (1986) 'Least-squares estimation of motion parameters from 3-D point correspondences,' *IEEE Conf. on Computer Vision and Pattern Recognition,* Miami, FL, pp. 198-201.

Kuperman, I.B., (1971) **Approximate Linear Algebraic Equations,** Van Nostrand Reinholt, London.

MacMillan, W.D., (1936) **Theoretical Mechanics. Vol 3: Dynamics of Rigid Bodies,** McGraw-Hill, New York, pp. 166-172.

Papoulis, A., (1965) **Probability, Random Variables, and Stochastic Processes,** McGraw-Hill, New York.

Paul, R.P., (1981) **Mathematics, Programming, and Control,** MIT Press, Cambridge, MA, pp. 29-32.

CHAPTER 11

Towards a Theory of Motion Understanding in Man and Machine

John K. Tsotsos
David J. Fleet
Allan D. Jepson

11.1 Introduction

The design of a system that understands visual motion, whether biological or machine, must adhere to certain constraints. These constraints include the types and numbers of available processors, how those processors are arranged, the nature of the task, as well as the characteristics of the input itself. This chapter examines some of these constraints, and in two parts, presents a framework for research in this area. The first part, Section 11.2, involves time complexity arguments demonstrating that the common attack to this problem, namely, an approach that is spatially parallel (as least conceptually), with temporal considerations strictly subsequent to the spatial ones, cannot possibly succeed. The essence of this claim is not a new one, and was motivated by similar comments by Neisser (1967), among others. What Neisser and others did not do, however, is provide a framework that is *plausible*. Expanding on the time complexity argument, we show that in addition to spatial parallelism, the basic system elements include hierarchical organization through abstraction of both prototypical visual knowledge as well as early representations of the input, and the separation of input measurements into logical feature maps.

The second part, Section 11.3, gives details for a specific component of the framework, namely that of the early organization and measurement of visual information, with most consideration given to velocity measurement. In particular, we concentrate on the very first functional level of visual processing. It is generally acknowledged that this level is responsible for the construction of an early representation so that certain primitive image properties are available to facilitate interpretation in subsequent levels. At present it is not known precisely what primitives are necessary and/or sufficient for interpretation. Even basic interpretation tasks, such as the identification of "meaningful edges" or "coherent regions" have proven to be surprisingly difficult. This is due, in part, to the use of a restricted set of primitives. It is our view that a richer representation of image structure can and should be generated by the first level. We suggest the extraction of primitives spanning several modalities including form, motion, and depth. Subsequent interpretation tasks, such as preliminary segmentation, may then use the different modalities simultaneously.

In constructing the early representation, it is important that the assumptions underlying the processing are stated explicitly so that we may distinguish between measurement and inference. We feel it is natural to restrict the first level to measurement processes only, thereby avoiding overly restrictive assumptions and any need for previous or concurrent interpretation. For example, the extraction of motion information should not require the identification of specific spatial features as is the case with current token-matching schemes (Ullman, 1979; Barnard and Thompson, 1980; Dreschler and Nagel, 1982). Similarly, we view retention of only limited information based on thresholds or maximum responses (perhaps from a family of channels) as a stage of interpretation and thus also beyond the initial stages of measurement. Such is the case with current gradient-based approaches to motion extraction in which unique spatial and temporal gradients are chosen at various locations (usually only those with large amplitudes or special properties) and used to solve some form of the motion constraint equation (Limb and Murphy, 1975; Horn and Schunck,

1981; Nagel, 1983).

Following the discussion on measurement in general, we consider the measurement of directional information, i.e., local orientation and velocity. In particular, we present several tools within a hierarchical computational framework which can be used to construct mechanisms that are local in space-time and selectively tuned to narrow ranges of orientation and speed. The hierarchical structure is essentially a parallel processing scheme in which nodes need compute only a weighted sum of inputs from within a small spatio-temporal neighborhood of nodes at the previous level, and/or from neighboring units at the same level. This type of processing is plausible for neural networks and parallel machines, moreover, it leads to an efficient computational scheme on serial machines. The scheme is easily and completely analyzed with the use of Fourier analysis. In the final section we discuss the current approach in relation to orientation and motion sensing in biological systems, and contrast it with current machine vision approaches to motion extraction.

11.2 The Time Complexity of Visual Perception

11.2.1 The Role of Time in Vision

Motivation

Biological sensory systems are examples of *real-time* systems, i.e., time-varying input is applied at the sensor and time-varying actions are produced such that an examiner can verify whether the input was understood, usually with no significant delay. The visual world is a spatio-temporal one, and perception takes place within a spatio-temporal context. In the past most computer vision research has dealt with static images, an unnatural kind of input since biological visual systems are almost never presented with a single time slice of the visual world out of its spatial and temporal contexts. For example, some current connectionist models of visual perception, particularly those claiming biological motivation, lack serious consideration of temporal issues, suggesting that the temporal issues are insignificant or that the problems raised by temporal issues can be simply solved (Ballard, 1985). This chapter puts forth a contrary position: the temporal issues are important and complex. Moreover, their consideration is crucial to a model if claims of biological plausibility are made.

There are three key ways in which time plays an important role in visual information processing:

- the world, and thus visual input, is inherently time-varying;

- the perceptual response in humans has an important temporal component; and

- computation requires time, with processors, as well as memory, having limited capacities, thus the computational demands of the task must be seriously examined.

Time does not even enter the picture for most researchers if a motion analysis task is not under consideration. The nature of the computation is not seriously considered either, and most researchers seem to believe that if a process is slow, then parallelism or a speed-up of processors will solve the problem. In addition, a tacit assumption is present in most past work on motion analysis and understanding. It has been assumed that techniques that have been used for the analysis of static images are directly relevant for the time-varying case, and that temporal processing strictly follows spatial analysis. We suggest that this is not necessarily the case. We claim that, generally, time and space are inseparable aspects of visual information processing. Finally, perhaps the most important observation is that algorithms for machine vision must be formally *time-bounded*. This is critical if real-time performance is expected. This issue of time bounds for vision algorithms has been given very little attention in the literature.

The strong coupling of space and time is not a new idea, and has been bounced around other scientific communities for quite some time. Perhaps the most famous instance of this controversy is due to Albert

Einstein. At a 1922 meeting of the French Philosophical Society, an eminent philosopher, Emile Meyerson, asked Einstein: "Is spatialization of time, i.e., the tendency to regard time, 'the fourth dimension', as not being essentially different from the spatial dimension, a legitimate interpretation of Minkowski's fusion of space and time?" [The interested reader can find several interesting examples of the static interpretation of space-time that were current in 1922 in (Capek, 1981).] Einstein's reply was very brief and simple, "It is certain that in the four dimensional continuum all dimensions are not equivalent". Einstein was not terribly interested in this issue. He did, however, make several other statements regarding space-time, such as "we cannot send wire messages to the past", and "the becoming in the three dimensional space is somehow converted to a being in the world of of four dimensions". Considered carefully, these statements are contradictory, since the latter implies a static interpretation while the former does not. This story is quite a bit longer, and includes many other philosophers and scientists. Including the story in this chapter in its entirety would be difficult, and it would not serve much purpose. The following point, however, is clear: the problem of how to deal with the interpretation of space-time has a deep and broad nature, is exceedingly difficult, and has occupied the best minds through history. Simple solutions within artificial intelligence, which are geared towards the solution of immediate tasks are unlikely to address most of the deep issues that are raised by space-time. Since virtually all past attempts within artificial intelligence have implicitly assumed the static interpretation of space-time, and have been unsatisfactory, this chapter will put forward the position that the other interpretation must be tried, and will outline part of a long term research effort in our laboratory to develop a computational model of visual perception in which space and time are inextricably linked in representation and computation.

The Temporal Characteristics of Perception

There is a great deal of research in the biological sciences that addresses the temporal aspects of physiological and perceptual responses. Good overviews of this research can be found in (Ganz, 1975; Uttal, 1981; Watson and Ahumada, 1985; Nakayama, 1985) among other sources. Some of the relevant conclusions are summarized in the following statements.

a) Reaction time is a composite delay consisting of at least the latencies of individual neurons, conduction time along neural pathways and processing time.

b) The progressive prolongation of the neural response at each level, contributes to a variety of visual phenomena that involve temporal interactions.

c) The response to a visual stimulus outlasts the stimulus by a considerable time, so that traces of two non-overlapping physical events can exist simultaneously within the perceptual system.

d) It seems that perceptual events have minimum duration of about 100 ms.

e) A given neuron has a response time of between 1 to 10 ms, but the system has a temporal integration time of 10 - 100 ms or more. That is, although individual neurons can respond quickly, in order for the stimulus to be processed by the entire system and for the system to reach a 'conclusion', much more time is required. The critical duration for perfect temporal integration is 100 ms, and arises from Bloch's Law (Ganz, 1975). It is important to note that the critical duration is defined as the time period during which there is no 'interfering' stimulus, and not as the duration of the stimulus, which may be much shorter. For specific tasks, the temporal integration time is found to be even longer. For direction discrimination tasks, at threshold contrast, 410 ms are required (Watson, et al., 1980), while for speed difference detection tasks above threshold contrast, 200 ms seem to be required (McKee, 1981). Other estimates suggest that brightness discrimination tasks require 50 - 100 ms of temporal integration, while form discrimination, including letter recognition tasks, require 200 ms or longer (Estes, 1978).

g) An upper limit on the number of simple, consecutive items that can be recognized in a stream is 10 per second (Potter, 1975). In other words, at least 100 ms of computation time is required for a given stimulus, and this is consistent with the critical duration for temporal integration.

Temporal considerations lead to the conclusion that an abstraction mechanism is a necessary component of perceptual models. There is an inherent limitation due to the nature of biological hardware. Neurons can maintain information, i.e., remember it, only for a short period of time, and in some works, this effect is termed iconic memory (Sperling, 1963). How long they can maintain a given signal is a function of neural characteristics, including rise time, fall time and degree of temporalsmoothing, and what is happening at the neural synapses. In other words, the temporal persistence of information is time-limited. As long as a given neuron can hold a signal, another neuron can read it. The act of reading the signal takes time, using up some of the time of the signal's persistence. While one neuron reads a signal, the input for the other neuron may be changing. If information comes into the system too quickly, that is, persistence is small, processing units do not have sufficient time to read all the information necessary for processing. Spatial abstraction thus permits larger spatial chunks of information to be read at a time, and temporal abstraction permits longer temporal chunks of information to be read.

11.2.2 The Nature of the Computational Problem

What follows is a simple demonstration that the task of visual motion understanding is very demanding computationally, and that sophisticated tools are necessary in order to cope with the demands. The discussion was motivated by comments by Neisser (1967) among others, that a spatially, parallel model of perception is inadequate quantitatively, simply because the brain is not large enough. We will take this one step further, and will attempt to answer the question, "*If spatial parallelism is insufficient, what are the characteristics of a sufficient mechanism?*" The argument is only summarized here and is given in detail in (Tsotsos, 1988).

The argument will take the form of a time complexity discussion, attempting to model the overall performance of an entire vision system. The main question that will be addressed is, how much processing must be done versus how fast the processing system operates and how much time it has in order to operate? Of interest is, how much can a visual processor made up of elements with neural speeds in a bottom-up, single pass manner without any a priori information, actually compute in 100 ms? Note that this argument has the nature of a "back of the envelope calculation". If we believe that vision can be modelled computationally, then such an exercise can tell us what size of problem is actually being solved. It should be noted that it is proven that some simple vision problems are NP-complete, e.g., labelling of polyhedra (Kiroussis and Papadimitriou, 1985).

Let us define a visual processing system for the purposes of this discussion. Assume that the sampled input is represented retinotopically, i.e., as a two-dimensional array, with P elements within the array. At each element, M measurements are taken. These measurements may include intensity, wavelength, binocular disparity, and a temporal gradient for each of these. The visual prototypes are stored in a knowledge base, KB, using no particular representation, and there are NP prototypes. Each prototype contains all the information necessary for the recognition of a prototype, including its variations, as well as a label. A visual processor is defined that can select a collection of array elements, their associated measurements and one prototype, and can determine whether or not the collection of measurements over the selection of locations optimally contains an image-specific projection of the prototype. This is clearly not a simple task. The KB may be thought of as containing specific image intensity patterns or complex symbolic scene entities. It does not really matter for our purposes, but this does point out that the argument is equally applicable to "early" and "high level" vision tasks. Let us call the collection of array elements with their corresponding measurements a "visual field".

This processor requires PS seconds to complete this task, i.e., select two inputs, match and compute a response. There are potentially $2^{P \times M}$ collections of measurement/location pairs. A time complexity function for an algorithm expresses its time requirements by giving for each possible input length, the largest amount of time needed, in terms of the input length. Thus, the time complexity function for this visual processor is

$$2^{P \times M} \times NP. \qquad (1)$$

A problem is intractable if no polynomial time algorithm can possibly solve it, and an algorithm is said to solve a problem if it can be applied to any instance of the problem, of any size, and the algorithm is guaranteed to produce a solution. The above expression shows that the processing as defined is intractable. In fact, a priori, there is no way to predict which visual field will represent an image-specific projection of a given prototype. Note that this expression counts data items that must be processed and does not enumerate images. Of course, vision is an effortless sensation for humans, and this result is not terribly insightful. Parallelism alone is not the answer.

With the most simple algorithm, a single processor must consider, in the worst case, each visual field against each stored prototype, and in the average case if instances of the prototypes occur with equal probabilities, half that number of comparisons. Neisser pointed out that even adding parallelism would be insufficient to deal with the computational demands of this simple model. It should be obvious that a parallel scheme requires serious consideration of the problems of communication, shared resources, synchronization, task scheduling, etc.. We will ignore these problems and assume that they can be resolved - there would be no impact on the model and the results claimed in this paper. PP stands for the effective processor speed-up due to parallelism and it is surely the case that in order to add a speed-up factor of one, many processors must be added to the system. Given the fact that there are perhaps 10^8 neurons in V1 of the visual cortex (Cowey, 1979), and several times that in the entire visual cortex, then the effective speed-up due to parallelism must be much less than this figure.

For the remainder of the paper we will be addressing the question: How does the human visual system manage to inspect the entire visual field and recognize, i.e. label, a scene in 100 ms of computation? Given PP degree of parallelism and PS as the time it takes for a single processor to complete one operation, the time constraint on the computation time of the visual processor is

$$100ms \geq \frac{2^{P \times M} \times NP \times PS}{PP}. \tag{2}$$

If P is the number of photoreceptors in the retina (130,000,000), M is 6 and NP is 10,000,000, then PP is approximately $10^{234000006}$ with only the final digit of the exponent being due to the terms other than the number of visual fields. This is simply impossible for human vision regardless of the value of PS since neural speeds are in the 1 - 100 ms range.

Clearly the algorithm must be made more efficient. Efficiency can be gained by attacking the knowledge base search through a process of successive refinement. Assume that we can build a binary tree whose leaves are the prototypes of the knowledge base, and whose nodes are super-classes of prototypes. This is similar in some respects to building a semantic network knowledge base. There is no real reason for insisting that the tree be binary. The key is that a tree search will reduce the search through the collection of prototypes logarithmically. The processing system speed is now

$$100ms \geq \frac{2^{P \times M} \times \log_2 NP \times PS}{PP}. \tag{3}$$

This clearly is insufficient, and thus the standard technique employed by all knowledge-based vision systems, namely prototype organization, is at best a small contributor to defeating the complexity problem of vision.

Further economy can be gained by noting an important physical constraint: the physical world is spatio-temporally localized. Thus, it is not the case that all 2^P visual fields are reasonable to consider. In fact, since objects are not spread arbitrarily in 3-space, and events are not spread arbitrarily in the time dimension, the major source of complexity is significantly reduced. Assuming a hexagonal image, and that only hexagonal, contiguous groupings of array elements are considered as visual fields, then some simple geometry yields

$$\frac{P^{1.5}}{3\sqrt{3}} + \frac{P}{2} + \frac{5\sqrt{P/3}}{8} \tag{4}$$

as the number of visual fields that need be considered. This reduces the estimate for PP in the expression,

$$100ms \geq \frac{\left[\dfrac{P^{1.5}}{3\sqrt{3}} + \dfrac{P}{2} + \dfrac{5\sqrt{P/3}}{8}\right] \times 2^M \times \log_2 NP \times PS}{PP} \tag{5}$$

and PP is estimated at 8.4×10^{14}. At least the number of zeroes will now fit on a page!

Another physical constraint can be applied, namely that the interpretation of visual input does not necessarily involve all visual measurements. For example, measurements of temporal characteristics for static scenes are not useful. If it were possible to quickly determine which types of measurements are relevant for given visual input, then further reduction in computational expense is possible. Let μ represent the number of measurements that are relevant for a given input. Thus, the number of possible subsets of measurements is $2^\mu - 1$. The null set is not included because it is known that given the response some non-null subset must be the right one, and the figure is now sufficiently small so that the null set from the count will make a difference. This could be implemented via a computation of "pooled response". If the pool of measurements of a given type are significantly non-zero, then a separate signal is produced identifying the measurement. A direct result is the segregation of measurements into separate maps - an idea that arose in "intrinsic image" theory (Barrow and Tenenbaum, 1978) and in "feature integration" theory (Treisman and Schmidt, 1982). It is currently believed that there are several *maps* in the visual cortex, organized retinotopically, with each map having its own specific mapping characteristics, seemingly different from the other maps. It has been claimed that there are between 15 and 20 of these physical maps, (van Essen and Maunsell, 1983), and that each map contains significant populations of different types of neurons. Thus, there are many more times this number of *logical feature maps*. We view a logical feature map as an array of measurements along some particular dimension, such as normal velocity, stereo disparity, opponent color, orientation, etc.. A hierarchical organization has been described for retinotopic visual maps (van Essen and Maunsell, 1983) with organization such that the more abstract the information represented in a map and the smaller the neural extent of the map, the larger its receptive field. The new expression for processing time is

$$100ms \geq \frac{\left[\dfrac{P^{1.5}}{3\sqrt{3}} + \dfrac{P}{2} + \dfrac{5\sqrt{P/3}}{8}\right] \times (2^\mu - 1) \times \log_2 NP \times PS}{PP}. \qquad (6)$$

Even for $\mu = 1$, PP is 1.3×10^{13}.

Unfortunately, P is still too large. Further reduction in P would mean reducing the resolution of the visual image, and simultaneously, abstracting the input in order to maintain its semantic content. Let Π be the size of the new abstracted array. What size of array leads to complete inspection within the time constraint? The processing time expression is

$$100ms \geq \frac{\left[\frac{\Pi^{1.5}}{3\sqrt{3}} + \frac{\Pi}{2} + \frac{5\sqrt{\Pi/3}}{8}\right] \times (2^{\mu} - 1) \times \log_2 NP \times PS}{PP}. \qquad (7)$$

For values of PP within biological ranges, i.e. less than 10^8, $\mu = 1$, namely, for the simplest input possible, and PS = 100 ms, it is found that at for PP $= 10^6$, $\Pi = 2269$, and for PP = 1000, $\Pi = 37$ Details and further experimentation can be found in (Tsotsos, 1988). It is known that the number of hypercolumns in area V1 of the virtual cortex is in the 5000 - 6000 range (Scholl, 1956), and each of these may be thought of as an array element of the visual map. Thus, 6000 seems to be an upper bound on the size of the maps, and it is likely that the maps are smaller. By observation from the diagrams in (van Essen and Maunsell, 1983), the reduction in neuronal area from V1 to MT or V4 may be as large as 0.1. The figures derived above for the size of the array that can be entirely inspected within 100 ms are consistent with these two anatomical observations.

To summarize this section, by using only time complexity arguments, it has been demonstrated that human vision must develop rather specialized structures in order to cope with the combinatorics involved in the interpretation of visual input. A plausible combination of specialized structures consists of:

a) parallel processing, with an effective speed-up of significantly less than 10^8 ;

b) prototype organization, reducing search time at least logarithmically.

c) reduction through abstraction of the size of the input map so that there are relatively few array elements, i.e., less than 3000, to combine into visual fields for matching; and

d) computation of separate measurement maps each with an associated pooled response indicating whether or not a particular map contains important input.

11.2.3 Implications

The implications of the three basic ways in which time complexity considerations affect visual processing models are far reaching:

1) An obvious implication is that there are no units that compute static information, divorced from time. Even stationary edges exist during a time interval. Common problems of noise sensitivity, smoothing and ambiguities faced by all vision algorithms can be ameliorated by the effect of temporal integration and the constraint of local temporal continuity. Motion and form computations, motion and depth computations, or motion and color computations are not necessarily performed independently. The fact that most visual areas seem to contain neurons computing different types of information, e.g., opponent color and orientation selectivity (Cowey, 1979), leads one to hypothesize that communication and cooperation among these differing units is a necessity for correct computation.

2) A second implication is the important use of abstraction both for feature maps and for prototype organization. We note that both temporal and spatial abstraction is necessary. When modelling spatial abstraction, where the spatial input is presented in parallel, hierarchies of units can be constructed, with each level looking at progressively larger spatial extents, computing more and more abstract entities. This has been borne out by experimentation with the receptive field properties and hierarchical organization of the various visual cortical areas (Gross, 1973). What is the analogy in time? Temporal information is presented to the system serially, In order for a single unit to look at progressively larger temporal extents, it must integrate over those temporal extents. Units with input that exists

over a larger temporal extent, compute abstractions that reflect longer events. Note that it is probably desirable to have many pathways of differing speeds, since it is not possible for a single temporal scale to optimally capture all temporal extents. This is exactly the same argument as for multiple spatial scales, and has appeared in that context in several works. Interestingly enough, at least three types of pathways have been found that can be characterized by the speed of their response, the so-called X, Y, and W pathways (Stone, et al., 1979). Importantly, units from different pathways may combine, and then the issue of synchronization and semantic meaningfulness must be considered.

3) Since the visual world is *spatio-temporal,* and not purely spatial or temporal, *all* units involved in abstraction must necessarily involve both spatial and temporal abstraction. Units may be *selective* for certain combinations of inputs, e.g., response is high for small ranges of values in the combination and drops greatly if the inputs are outside these tight bounds, or they may be *sensitive* to some combinations of inputs, e.g., response increases monotonically with increasing stimulation by a certain combination of inputs. Thus, individual units may incorporate varying degrees of spatial and tmporal selectivity and sensitivity. Single unit spatio-temporal abstraction implies:

a) there must be a semantically meaningful relationship in space and in time among all input units; and

b) the bottom-up inputs must be *synchronized,* i.e., must be derived from the same spatio-temporal input *wave.* Lateral and top-down inputs must also conform to similar constraints. The flow of data is constrained by temporal integration times at all units, and thus it is not necessarily true that each unit completes processing, passes on information to the next level, and then takes on new input. Rather, it seems that the entire network is busy at work on several time slices of events, so that neural elements are not merely computation steps, but more correctly, introduce *delay* into processing with continuous computation.

4) Parallelism and knowledge base organization are only minor players among the tools required to beat the time complexity of visual perception. Much greater gains are made by abstracting the input into small, separate maps of measurements.

5) Only a coarse level of analysis is possible within 100 ms. Any detailed examination of image characteristics must involve an attentional capacity. That discussion is the topic of future research.

Our conclusions may in fact lead to an inversion of one of the accepted paradigms in computer vision, namely, visual modules compute specific visual entities roughly independently of one another, and then information is combined. We suggest that since all neural computing elements are necessarily spatio-temporal, then a more plausible mechanism for early vision is that each level of computation *specializes* the abstraction computed at the previous level, and further at each level there may be several different specializations, occurring in parallel. Even combinations which occur must do so in a synchronized manner, and can be regarded as specializations since in effect two separate responses intersect. Many of the levels may not not have immediately obvious semantics, however, since arbitrary decompositions may be necessary in order to achieve as efficient an organization as possible from a search point of view. In fact, if a bit of further speculation is permitted, this may be the reason why it is so difficult to ascribe semantics to the characteristics of some individual neural responses - they simply may not have intuitive functional roles and their presence may be primarily for efficiency reasons. The responses that are useful are the ones that are the most specialized possible.

11.3 Measurement and Hierarchical Representations in Early Vision

11.3.1 What is Measurement?

In discussing early visual processes it is necessary to distinguish between measurement and inferential processes. A measurement is obtained by ascertaining the amount of a particular quantity in terms of standard units. To say that a given process is a measurement we need to state precisely what quantity is being measured, and what units we are using to represent this quantity. The latter requirement is to eliminate subjective criteria such as "strong" or "weak". By contrast, in making an inference one arrives at an opinion, or comes to accept a probability, on the basis of premises and available evidence. The evidence may be incomplete, the premises may include various assumptions, and the inferences derived may be incorrect. A single process can often be viewed as either a measurement or an inferential process. The distinction often lies with the way the output is interpreted. For example, if certain measurements are always ignored or discarded, as is the case when only the largest of the responses from several operators is retained, then there are certain tacit assumptions. For the purposes of this chapter, we call a process inferential if it can be naturally thought of as arriving at its result on the basis of various assumptions. Measurement processes, on the other hand, do not require *a priori* assumptions of the input signal. We call these processes "blind" to emphasize their non-interpretive nature.

It is generally acknowledged that the first functional level of visual processing is responsible for the construction of a representation which makes various visual primitives, such as orientation and velocity information, explicit. The next level of processing is expected to use this representation to infer more complicated entities, such as texture or surface descriptions. We propose to limit the very first functional level to consist of only measurement processes. The primary reason is that we wish to provide subsequent inferential processes with a complete representation of the raw image in which primitives of various modalities, e.g., form and motion information, are available. The inferential processes can then use these different modalities simultaneously, which we feel will facilitate the inference process and increase robustness. Furthermore, since the initial representation is complete and obtained by measurements, it can be used without concern for any restrictive assumptions on which it might have otherwise been based (Jepson and Fleet, 1987).

In the remainder of this chapter we discuss how orientation and velocity information can be measured, and we discuss the theoretical limits of a simple class of operators for doing this. Then in Section 11.3.3. and Section 11.3.4. we present an efficient scheme for the implementation of suitable operators, and present examples. Finally, in Section 11.3.5. and Section 11.3.6. we consider the implications of our approach for some of the current theories of early visual processing in man and machine.

11.3.2 Directional Information and its Measurement

Both the orientation of simple spatial patterns and their normal velocities may serve as generic primitives for use in subsequent stages of processing, and they can be measured to within particular theoretical limits. In this section we first state precisely what is being measured in each case. This statement is done by appealing to a frequency space representation of the primitive image properties. From this formulation we obtain design criteria for some simple measurement processes. Finally, we briefly discuss constraints on the accuracy obtainable with these techniques.

Orientation Information

An oriented, straight edged image component, e.g. a step edge, will have its signal energy concentrated along a line through the origin in frequency space. For example, consider an ideal case in which the image intensity function is a Dirac delta function distributed along a line $x_2 - mx_1 = 0$, namely,

$$I(x_1, x_2) \; = \; \delta(x_2 - mx_1) \; . \tag{8a}$$

This has the Fourier transform,

$$\hat{I}(k_1, k_2) \; = \; \delta(k_1 + mk_2) \; . \tag{8b}$$

The slope of the line in the frequency domain depends continuously on the orientation in the image. Therefore, to measure orientation information near a given spatial location, we need to measure the signal energy (within a spatial window) concentrated about lines through the origin in frequency space. To do this we can use filters selectively tuned to a narrow range of spatial orientations, in particular, if a given filter is to have the preferred orientation m then its amplitude spectrum must be concentrated about a line through the origin with slope $-1/m$.

Velocity Information

Image velocity also has a simple representation in frequency space [For a more detailed review see (Fleet and Jepson, 1984,1985a; Adelson and Bergen, 1985; Watson and Ahumada, 1985)]. Consider the two-dimensional intensity function, $I_o(\vec{x})$, translating with a constant velocity, $\vec{v}$. The corresponding spatio-temporal intensity function is given by

$$I(\vec{x}, t) \; = \; I_o(\vec{x} - \vec{v}t) \; . \tag{9a}$$

The Fourier transform of Eq. 9a, is

$$\hat{I}(\vec{k}, \omega) \; = \; \hat{I}_o(\vec{k}) \; \delta(\omega + \vec{v} \cdot \vec{k}) \; , \tag{9b}$$

where $\hat{I}_o(\vec{k})$ is the transform of the static spatial pattern, $I_o(\vec{x})$. Here, $\vec{x}$ is the vector, (x_1, x_2), and $\vec{v} \cdot \vec{k}$ denotes the dot product. Note that $\hat{I}(\vec{k}, \omega)$ is the non-orthogonal projection, parallel to the ω-axis, of $\hat{I}_o(\vec{k})$ onto the plane, $\omega = -\vec{v} \cdot \vec{k}$. The speed, $|\vec{v}|$, determines the angle between the two planes, $\omega = -\vec{v} \cdot \vec{k}$ and $\omega = 0$. The direction of $\vec{v}$ determines the orientation of the plane about the ω-axis. Finally, there is a one-to-one correspondence between finite two-dimensional image velocities, $\vec{v}$, and the planes intersecting the origin in frequency space that do not contain the entire ω-axis.

Pure image translation has a simple representation as a plane in frequency space for which simple filters may be constructed. Moreover, such filters may be applied successfully to relatively large spatio-temporal windows in certain domain specific applications. For general utility, however, we desire only local translational information because perspective projection, non-rigid motions, motion in three-dimensional, and rotations all combine to produce varying profile two-dimensional images over any but relatively small image regions. Within these local windows spatial intensity changes will most often appear straight edged and thus we can only expect to measure velocities normal to local orientation. This aspect of motion measurement has been called the *aperture problem* (Marr and Ullman, 1981). Thus we are led to consider the measurement of local two-dimensional normal velocity.

If the two-dimensional image velocity is $\vec{v}$ and the normal is $\vec{n}$, then the normal velocity is

$$\vec{v}_n = (\vec{n} \cdot \vec{v})\, \vec{n}. \tag{10}$$

The non-zero Fourier components of an oriented pattern moving with normal velocity, $\vec{v}_n$, lie along the line given by (see Figure 1)

$$(\vec{k}, \omega) = c(\vec{v}_n, -|\vec{v}_n|^2), \quad c \in \mathbb{R}. \tag{11}$$

This line is contained in the plane obtained from the two-dimensional image velocities $\vec{v}$, and it passes through the origin. The slope of the line, relative to the plane $\omega=0$, corresponds to speed, and the direction of motion determines the orientation of the line about the ω-axis, relative to a fixed spatial axis. In general, there is a one-to-one correspondence between lines intersecting the origin in frequency space and two-dimensional normal velocities, $\vec{v}_n$.

Design Criteria

We consider only the simplest processes capable of measuring directional information. In particular we need only consider linear operators of the form,

$$R(\vec{x}_o) = \int_{-\infty}^{+\infty} \int_{-\infty}^{+\infty} K(\vec{x} - \vec{x}_o)\, I(\vec{x})\, d\vec{x}. \tag{12}$$

In accordance with the conditions outlined above we must constrain such operators to be local while maintaining some degree of directional selectivity. In order to ensure that the response, $R(\vec{x}_o)$, depends mainly on the local structure of the image, $I(\vec{x})$, we assume that the magnitude of the impulse response function, $|K(\vec{x})|$, is small for large values of $|\vec{x}|$. In particular, we assume $K(\vec{x}) = K_o(\vec{x})\, W(\vec{x})$ for some bounded function, $K_o(\vec{x})$, where $W(\vec{x})$ is a windowing function. Then, the application of $K(\vec{x})$ can be thought of as a linear operation applied to the windowed input, that is,

$$R(\vec{x}_o) = \int_{-\infty}^{+\infty} \int_{-\infty}^{+\infty} K_o(\vec{x} - \vec{x}_o)\, I_w(\vec{x}\,;\vec{x}_o)\, d\vec{x} \tag{13}$$

where

$$I_w(\vec{x}\,;\vec{x}_o) \equiv W(\vec{x} - \vec{x}_o)\, I(\vec{x})\,. \tag{14}$$

We may now formulate our objective in terms of two conditions which, in general, should be satisfied by a filter which is selectively sensitive to local orientation or two-dimensional normal velocity.

1) *Localization in Space-Time.* The extent of the impulse response, or support, should be restricted to a relatively narrow aperture in space-time.

2) *Orientation/Velocity Specificity.* The filter should be tunable to a narrow range of orientation or normal velocity, with its amplitude spectrum concentrated about a line through the origin in frequency space.

For example, for velocity specificity, the amplitude spectrum should fall mainly within a cone, $C(\vec{v}_n, \theta)$, about a line with a relatively small opening angle, θ (see Figure 1).

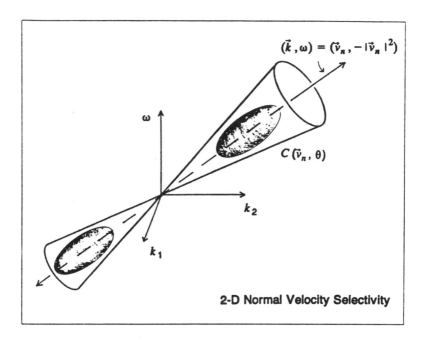

Figure 1. The cone shows the region in frequency space corresponding to a narrow range of two-dimensional normal velocity. The shaded region illustrates the dominant part of the amplitude spectrum of a filter selectively sensitive to such a range.

There are several important theoretical restrictions on the use and satisfaction of these two conditions. These are rather complicated and are discussed only briefly below. We refer the interested reader to Jepson and Fleet (1987) for elaboration. Here, it suffices to draw attention to the main issues. First, localization is meaningful only in relation to the scale at which one-dimensional orientation and normal velocity are reasonable approximations to local image structure. In effect, the practical usefulness of Condition 2 requires Condition 1. Second, the well-known uncertainty relation (Brillouin, 1956, p.89) imposes finite theoretical limits below which the two conditions cannot be satisfied simultaneously, as explained further below.

The Effect of Windowing

We need to examine the information that we can expect to measure in the windowed image, $I_w(\vec{x}, \vec{x}_o)$, given in Eq. 14. To do this, we note that a multiplicative windowing operation in space-time corresponds to the convolution of the window and the input signal in the frequency domain. For relatively smooth local windows, the result is a low-pass smoothing (i.e., blurring) of the input signal's Fourier transform. This effect is made clear by the following simple calculation for orientation information. Consider the windowed image given in Eq. 14, where, for convenience, take $\vec{x}_o = 0$ and let the window be a Gaussian with standard deviation σ_w. Then the Fourier transform of the windowed input is

$$\hat{I}_w(\vec{k}; 0) = \hat{G}(\vec{k}; \sigma_w) * \hat{I}(\vec{k}), \tag{15}$$

where $\hat{I}(\vec{k})$ is the transform of the image intensity. The blurring in frequency space, caused by the windowing operation, is apparent in Eq. 15, i.e., the smaller the window the greater the blurring.

Moreover, from this equation we can estimate the number of independent orientation channels that can be obtained through this window for a small range of frequencies, that is, for $|\vec{k}|$ near some f_o. A simple way to impose the independence of the channels is to arrange their amplitude spectra so that they do not overlap significantly. As a measure of the radius of the transform, we take one standard deviation of $\hat{G}$, i.e., $R \equiv 1/\sigma_w$ in angular frequency. The opening angle of the orientation cone, which just contains one standard deviation of $\hat{G}$ centered at $(f_o, 0)$, is given by (see Figure 2)

$$\theta = 2 \, arcsin(\frac{R}{f_o}). \tag{16}$$

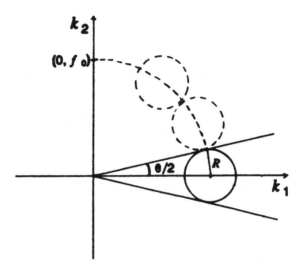

Figure 2. It is of interest to consider how many non-overlapping (in-dependent) orientation measurements can be made at a given scale (about f_0) within a given window size.

From this we find that the maximum number of orientation channels, centered at frequency f_o and separated by at least one standard deviation in frequency space, is

$$N = \lfloor \pi/\theta \rfloor, \tag{17}$$

where $\lfloor x \rfloor$ denotes the integer part of x. Note that it is sufficient to fill only 180°. For example, consider a frequency, f_o, such that the associated period, $2\pi/f_o$, is equal to the spatial diameter of the window, $2\sigma_w$. Then Eq. 17 provides $N = 4$. For higher frequencies it is possible to resolve more orientations, and in fact the number of independent orientation channels grows nearly linearly with the frequency. A similar argument shows that the number of independent normal velocity channels, that can be obtained through a given spatio-temporal window, grows quadratically with the frequency, $f_o \equiv | (\vec{k}, \omega) |$.

In summary, the localization/tuning trade-off is more acute for lower frequencies, and in general, higher frequencies should be used to obtain better resolution. In other words, the polar specificity of orientation and normal velocity, necessary for extraction, will not be intact for low frequency components in a local analysis due to the windowing. This suggests the use of radial band-pass properties with amplitude spectra having ellipsoidal shapes as shown in Figure 1. This constraint naturally complements a desire for multi-scale processing, since a family of band-pass mechanisms is inherently scale specific.

11.3.3 Hierarchical Processing

Given the constraints outlined above, a family of linear, directionally selective filters is easily constructed, each of which may be applied to the image. Unfortunately, the direct application of numerous filters as two-dimensional or three-dimensional discrete convolutions is inordinately expensive in both computation time and storage. For example, in the case of normal velocity extraction, filters must differ in scale, speed, and orientation selectivity. Obviously, if each of these parameters is discretized in n steps then there must be n^3 filters. In general, each will be inseparable and require m^3 multiplications for each spatio-temporal location, where m is the approximate width of support in space-time. We present an alternative approach in terms of a hierarchical computational framework which is more efficient and provides convenient tools for design and analysis.

The hierarchical computational structure is based on layers of linear processing units, a simple form of which is shown in Figure 3. The response of a unit is given by two parts. The α_j represent weights on the inputs from the previous layer, and the β_j represent a recursive filter with weights on the responses of neighboring units in the same layer. In general, the response of a unit at layer v is given by

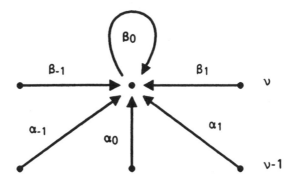

Figure 3. A simple cascade of explicit and implicit stages in one dimension.

$$\beta_0 \, R_\nu(\vec{x}, t) \;=\; \sum_{j=-p}^{p} \alpha_j R_{\nu-1}(\vec{x} + j\vec{\xi}_\alpha, \, t + j\tau_\alpha) \tag{18}$$

$$-\sum_{\substack{k=-q \\ k \neq 0}}^{q} \beta_k R_\nu(\vec{x} + k\vec{\xi}_\beta, \, t + k\tau_\beta) \,,$$

where $\vec{\xi}$ and τ are spatial and temporal offsets, $R_\nu(\vec{x}, t)$ denotes the response of a unit in layer ν, and $R_0(\vec{x}, t) = I(\vec{x}, t)$ is the input signal. In general, as is suggested from Figure 3., p and q will be small.

We may rewrite Eq. 18 in terms of convolution as follows:

$$B_\nu(\vec{x}, t) * R_\nu(\vec{x}, t) \;=\; A_\nu(\vec{x}, t) * R_{\nu-1}(\vec{x}, t) \tag{19}$$

where $A(\vec{x}, t)$ and $B(\vec{x}, t)$ are expressed in terms of Dirac delta functions, as follows,

$$A(\vec{x}, t) = \sum_{j=-p}^{p} \alpha_j \delta(\vec{x} + j\vec{\xi}_\alpha, \, t + j\tau_\alpha) , \text{ and} \qquad (20a)$$

$$B(\vec{x}, t) = \sum_{k=-q}^{q} \beta_k \delta(\vec{x} + k\vec{\xi}_\beta, \, t + k\tau_\beta) . \qquad (20b)$$

According to the convolution theorem, the convolution of two signals in space-time is equivalent to the product of their respective Fourier transforms. Therefore, the cumulative transform of the first L layers in the hierarchy is given by

$$\hat{H}_L(\vec{k}, \omega) = \prod_{l=1}^{L} \frac{\hat{A}_l(\vec{k}, \omega)}{\hat{B}_l(\vec{k}, \omega)} . \qquad (21)$$

To characterize the behavior of $H_L(\vec{x}, t)$ we may consider Eq. 21 directly, or we may consider the impulse response by taking its inverse Fourier transform. [For more details see (Fleet and Jepson, 1985a)].

11.3.4 Construction of Orientation or Velocity Selective Filters

In short, the present approach toward the extraction of directional information involves two main stages of processing. The first stage involves a scale specific decomposition. In other words, a family of radial band-pass filters is applied to the image, e.g., (Burt, 1981; Crowley, 1982). Each such filter will respond to all directional information over a finite range of spatial and temporal scales. The second stage is then used to specialize the outputs of the first stage, i.e., each scale specific filter, into channels tuned to narrow ranges of orientation or velocity. This is accomplished through the application of simple and efficient linear operations to the output of the first stage.

In the spatial case, (towards orientation tuning) we use sums and differences of outputs within a small spatial neighborhood, from the first level. This has the effect of selecting a stripe in frequency space (removing all but a stripe) from the isotropic band-pass spectrum of the initial level. With these filters the degree of specificity is controlled by adjusting the number of levels in the cascade and the form of processing at each level. Such an approach is easily extended to space-time with a three-dimensional, radial, band-pass filter for the initial stage, and local sums and differences of its output from within a small spatio-temporal neighborhood for the second stage. The sections on velocity tuning concentrate more on the differences in implementation and design necessitated by the temporal component. In time we require more efficient implementations which avoid the large amount of storage and computation required by a discrete FIR realization of temporal convolution. The surprising ease with which orientation and normal velocity are extracted from radial band-pass output will become clear below through examples.

In what follows we concentrate on the specialization of the scale specific response into directionally specific channels. Therefore, we may consider a single representative band-pass channel.

Initial Filtering: The DOG and CS Operators

For an initial stage of processing in the purely spatial case we use a difference of Gaussians (DOG) band-pass operator (Marr and Hildreth, 1980; Burt, 1981; Crowley, 1982),

$$DOG(\vec{x}) = G(\vec{x}, \sigma_c) - G(\vec{x}, \sigma_s) \, , \tag{22}$$

where $G(\vec{x}, \sigma)$ is a two-dimensional Gaussian with standard deviation σ. The amplitude spectrum for an instance of $DOG(\vec{x})$ is shown in Figure 4a. The DOG is not an essential quantitative component of the present approach but was chosen because of several appealing properties. It is isotropic, and the two-dimensional Gaussians are separable with good localization in space and the frequency domain simultaneously.

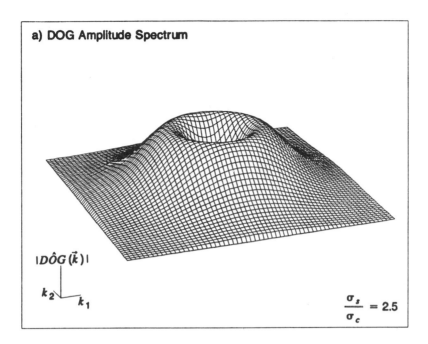

Figure 4a. A DOG amplitude spectrum.

As an initial stage of processing in space-time we borrow a center-surround (CS) operator which was originally derived as an extension of the DOG model to include temporal characteristics which agree qualitatively with the behavior exhibited by retinal units (Richter and Ullman, 1982; Fleet, et al., 1985). The extension includes exponential low-pass filters for both the center and the surround, with the surround delayed relative to the center. The impulse response function is written as follows:

$$CS(\vec{x}, t) = K(t; \lambda_c)\, G(\vec{x}, \sigma_c) - K(t-d; \lambda_s)\, G(\vec{x}, \sigma_s), \qquad (23)$$

where $G(\vec{x}, \sigma)$ is a two-dimensional, isotropic, spatial Gaussian, $d \geq 0$ is the surround delay, and,

$$K(t; \lambda) = \begin{cases} \lambda\, e^{-\lambda t} & \text{if } t \geq 0,\ \lambda > 0 \\ 0 & \text{if } t < 0 \end{cases} \qquad (24)$$

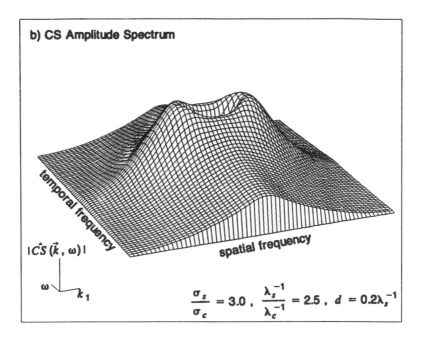

Figure 4b. The amplitude spectra for an instance of the CS filter. Given the radial isotropy of the spatial Gaussians it is sufficient to show the CS amplitude spectrum along one spatial frequency axis only; for convenience we let $k_2 = 0$.

is a temporal exponential low-pass filter with rise time λ^{-1}. Essentially, the parameters that affect the form of the model are the ratio of spatial standard deviations, σ_s/σ_c, the ratio of time constants, $\lambda_s^{-1}/\lambda_c^{-1}$, and the surround delay, d (Fleet and Jepson, 1985b).

The Fourier transform of Eq. 23 is

$$\hat{CS}(\vec{k}, \omega) = \hat{K}(\omega; \lambda_c)\, \hat{G}(\vec{k}; \sigma_c) - \hat{K}(\omega; \lambda_s)\, \hat{G}(\vec{k}; \sigma_s)\, e^{-id\omega}, \quad (25)$$

where $\hat{G}(\vec{k}; \sigma)$ is transform of the Gaussian, and $\hat{K}(\omega; \lambda)$ is given by

$$\hat{K}(\omega; \lambda) = \frac{\lambda}{\lambda + i\omega}. \quad (26)$$

$\hat{CS}(\vec{k}, \omega)$ is the difference of two three-dimensional low-pass filters. The form supported by the physiology has a radial band-pass amplitude spectrum which is illustrated in Figure 4b. Since two-dimensional normal velocities correspond to lines through the origin in frequency space, and since the CS amplitude spectrum is radially band-pass, the CS filter is sensitive to all two-dimensional velocities for a finite range of spatial and temporal scales. It is therefore a natural precursor to the extraction of velocity specific information, just as the DOG is for orientation selectivity. One convenient property of the exponential, Eq. 24, is that it may be realized as a simple difference, that is,

$$R(t) = K(t) * I(t) \tag{27}$$

can be computed as,

$$R(t) = \alpha R(t-1) + (1-\alpha) I_d(t) , \tag{28}$$

where $\alpha = e^{-\lambda}$, the rise time of the filter is λ^{-1}, and $I_d(t)$ is the discrete sampled input. This is more efficient than a discrete FIR convolution. We also exploit the separability of the Gaussians to realize the entire CS operator with $4m + 4$ multiplications, where m is the width of support.

Orientation Tuning Through Local One-Dimensional Sums

Our goal, given the DOG filter as a precursor, is to extract a stripe out of its band-pass spectrum such that the resulting spectrum falls within a cone (see Figure 1). We begin with the simplest case: the sum of three neighboring DOG outputs. This is given by

$$H(\vec{x}) = S(\vec{x}) * DOG(\vec{x}) \tag{29}$$

$$= \tfrac{1}{2}DOG(\vec{x}) + \tfrac{1}{4}DOG(\vec{x}+\vec{\xi}) + \tfrac{1}{4}DOG(\vec{x}-\vec{\xi}) ,$$

where $S(\vec{x})$ is defined using Dirac delta functions by

$$S(\vec{x}) \;=\; \tfrac{1}{2}\delta(\vec{x}) + \tfrac{1}{4}\delta(\vec{x}+\vec{\xi}) + \tfrac{1}{4}\delta(\vec{x}-\vec{\xi}) \;. \tag{30}$$

The corresponding Fourier transform is

$$\hat{H}(\vec{k}) \;=\; \hat{S}(\vec{k})\; D\hat{O}G(\vec{k}) \;, \tag{31}$$

where

$$\hat{S}(\vec{k}) \;=\; \tfrac{1}{2} + \tfrac{1}{4}e^{+i\vec{k}\cdot\vec{\xi}} + \tfrac{1}{4}e^{-i\vec{k}\cdot\vec{\xi}} \tag{32}$$

$$\;=\; \tfrac{1}{2}[1 + \cos(\vec{k}\cdot\vec{\xi})] \;.$$

The amplitude spectrum of the DOG filter is annular in shape (see Figure 4a). The amplitude spectrum of $S(\vec{x})$, equivalent to Eq. 32, is a two-dimensional oriented sinusoid. The crests and troughs of Eq. 32 are lines, orthogonal to $\vec{\xi}$, with values 1 and 0, respectively. One crest intersects the origin, $\vec{k}=\vec{0}$. The period of the wave is determined by the magnitude of $\vec{\xi}$. In particular, the period along the k_1 and k_2 axes depends solely on the values of ξ_1 and ξ_2, respectively.

Unfortunately, we find two problems with such a simple scheme: a ringing/specificity trade-off and a localization/specificity trade-off. In short, the use of small offset values, ξ_1 and ξ_2, in an attempt to increase localization and to avoid ringing, is accompanied by a loss of velocity specificity. Conversely, too much specificity results in an unacceptable degree of ringing in the amplitude spectrum, and a loss of localization information owing to larger offsets. An intermediate case is illustrated in Figure 5.

Multiple Levels and Denser Connections

Although the localization/specificity trade-off is, in part, theoretically unavoidable due to the uncertainty relation (see Section 11.3.2), there are several ways to avoid unwanted ringing while maintaining a reasonable degree of orientation specificity. One approach is to apply multiple layers

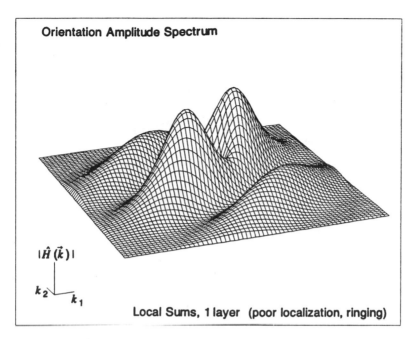

Figure 5. This plot illustrates the trade-off between specificity and ringing in the amplitude spectrum of an orientation selective mechanism constructed with only one layer of simple sums applied to the isotropic DOG. Larger offsets cause more ringing, and smaller offsets extract a broader stripe resulting in poor radial specificity.

of $S(\vec{x})$ in cascade. By adjusting the offsets at different levels the crests which cause unwanted ringing can be made to cancel with troughs from other levels. Figures 6a and 6b show the spectrum resulting from the application of two and three layers of local sums to the DOG output.

The use of multiple levels of simple sums does eliminate ringing while allowing for high degrees of specificity. The major drawbacks in their use are *i)* the added expense of each level, and *ii)* the loss of spatial localization. With such a simple scheme the localization/specificity trade-off is unacceptably severe and we should therefore attempt to obtain filters which are more near optimal. Essentially, the extraction of a slice from the isotropic band-pass spectrum amounts to low-pass smoothing in the direction of the preferred orientation. We are therefore led to consider the

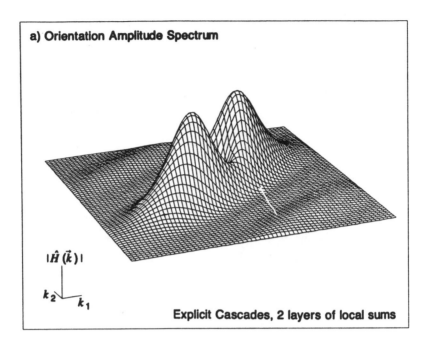

a) Orientation Amplitude Spectrum

$|\hat{H}(\vec{k})|$

k_2 k_1

Explicit Cascades, 2 layers of local sums

Figure 6a. The amplitude spectrum of a filter constructed with two layers of simple local sums. With the use of multiple layers of cascades we can obtain arbitrary finite degrees of velocity specificity with no appreciable ringing.

alignment of a greater number of local units, perhaps 5 or 7, with a smooth envelope on the weights. Natural choices for envelopes include a Gaussian window and a Kaiser window (Rabiner and Gold, 1975) since both exhibit good localization in space and frequency space simultaneously. Their analogues, the Gaussian and the spheroidal prolate functions, minimize different formulations of the uncertainty principle (Slepian, 1983). A relatively crude window, in terms of stop-band ripple, is sufficient since high frequencies have already been attenuated by the DOG.

Local Differences

In addition to enhancing specificity by narrowing the stripe through the origin along the preferred orientation, it is often desirable to shorten the frequency bandwidth of the resulting filter or to remove low frequency

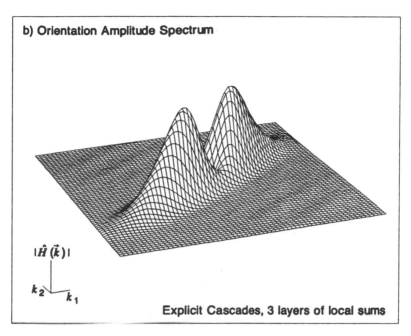

Figure 6b. An example with three levels of local sums.

information, especially that orthogonal to the preferred orientation. The latter is easily achieved by taking local differences in the direction perpendicular to that in which the sums were taken. For example, the difference operator,

$$D(\vec{x}) = \frac{1}{2}\delta(\vec{x} + \vec{\xi}) - \frac{1}{2}\delta(\vec{x} - \vec{\xi}) , \tag{33}$$

has the following Fourier transform,

$$\hat{D}(k_1, k_2) = i \sin(\vec{k} \cdot \vec{\xi}) . \tag{34}$$

The amplitude spectrum of Eq. 34 is periodic and zero (with discontinuities in the first derivative) on a line through the origin. The orientation of the zero lines is $\vec{\xi}$. An application of the difference operator in Eq. 33 to the result of two layers of local sums (Figure 6a) is shown in Figure 7.

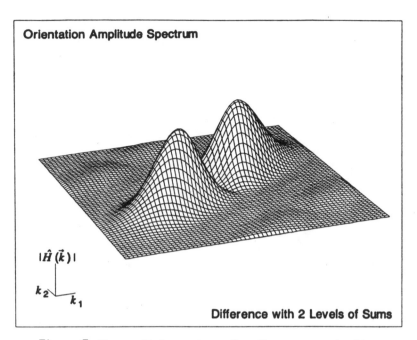

Figure 7. The amplitude spectrum of a filter constructed with a difference operator, Eq. 33, applied to the filter constructed with two layers of local sums shown in Figure 6a.

A somewhat wider region can be annihilated with a difference operator of the form,

$$D(\vec{x}) = \frac{1}{2}\delta(\vec{x}) - \frac{1}{4}\delta(\vec{x} + \vec{\xi}) - \frac{1}{4}\delta(\vec{x} - \vec{\xi}) \,. \tag{35}$$

The Fourier transform of Eq. 35,

$$\hat{D}(\vec{k}) = \frac{1}{2}(1 - \cos(\vec{k} \cdot \vec{\xi})) \,, \tag{36}$$

is more flat-bottomed in its troughs and will therefore attenuate a wider region about the line through the origin.

By applying $D(\vec{x})$ in the direction perpendicular to the direction in which the sums are taken we explicitly remove the sensitivity to orientations nearly orthogonal to the preferred orientation. This is significant because there is often a large amount of energy in low frequency components which can otherwise introduce unacceptable amounts of unwanted "noise" even after being attenuated by the local sum operations. It is also of interest to note that the local differences do not degrade localization noticeably since the offsets will generally be small and will only enlarge the impulse response in the direction orthogonal to the direction in which local sums were taken where localization is a problem.

Velocity Tuning Through Explicit Cascades

In the orientation case, a stripe was extracted in frequency space from the DOG spectrum. Here we show the extraction of a cylindrical region from the spheroidal CS spectrum. The major axis lies on a line through the origin, and the diameter determines the specificity, i.e., velocity tuning. The present approach is based on the ease with which a planar slice is extracted from the CS spectrum. The extraction of two orthogonal slices amounts to the extraction of the cylindrical region. Provided that the width of the slices is small relative to the low end of the spectrum already removed by the CS filter, the combined spectrum will fall mainly within a cone determined by the width of the slices.

Consider the sum of three local CS units (cf. Eq. 29),

$$H(\vec{x}, t) = S(\vec{x}, t) * CS(\vec{x}, t) \tag{37}$$

$$= \tfrac{1}{2}CS(\vec{x}, t) + \tfrac{1}{4}CS(\vec{x} + \vec{\xi}, t + \tau) + \tfrac{1}{4}CS(\vec{x} - \vec{\xi}, t - \tau),$$

where

$$S(\vec{x}, t) = \tfrac{1}{2}\delta(\vec{x}, t) + \tfrac{1}{4}\delta(\vec{x} + \vec{\xi}, t + \tau) + \tfrac{1}{4}\delta(\vec{x} - \vec{\xi}, t - \tau). \tag{38}$$

Here, $\vec{\xi} = (\xi_1, \xi_2)$ and τ are offsets in space and time, respectively. Of interest in the analysis of $H(\vec{x}, t)$ is the Fourier transform of Eq. 38 which is given by

$$\hat{S}(\vec{k}, \omega) = \tfrac{1}{2}(1 + \cos(\vec{k} \cdot \vec{\xi} + \omega\tau)) . \tag{39}$$

The amplitude spectrum of the CS filter is spheroidal in shape (see Figure 4b). The amplitude spectrum of $S(\vec{x}, t)$, equivalent to Eq. 39, is a three-dimensional sinusoidal plane wave. Its crests and troughs are planes, orthogonal to $(\vec{\xi}, \tau)$, with values 1 and 0, respectively (cf. Eq. 32), and one crest intersecting the origin. The period of the plane wave along each of the k_1, k_2, and ω axes depends solely on the offset values of ξ_1, ξ_2, and τ, respectively.

The specialization of the CS response into channels selectively sensitive to two-dimensional normal velocity can proceed in two stages as follows. In the first stage we apply an instance of the $S(\vec{x}, t)$ operator, with $\vec{\xi} = \vec{\xi}_1$ and $\tau = 0$. The resulting filter is sensitive to spatial features oriented according to $\vec{\xi}_1$ with no speed preference. The spatial and temporal scales are determined by the CS parameters. As the second stage of cascades, we again apply $S(\vec{x}, t)$, but with $\vec{\xi} = \vec{\xi}_2$ and $\tau = \tau_2$, such that, $\vec{\xi}_1 \cdot \vec{\xi}_2 = 0$. The resultant filter is specific to oriented spatial features, determined by the first stage, moving with a range of normal velocities (the center of which is determined by $|\vec{\xi}_2|/|\tau_2|$). Figure 8 shows three instances of speed selective filters (only one spatial dimension is shown). Assuming that the first level of cascades extracted orientations effectively parallel to $x_1 = 0$, Figure 8 shows filters sensitive to these orientations but differing in velocity specifically, with fast, moderate and slow speeds being selected in Figures 8a, 8b and 8c, respectively.

Low-Pass Temporal Smoothing and Other Cascades

Again there is a ringing/specificity trade-off if only one layer of local sums is used. Moreover, the trade-off is more severe for filters tuned to relatively slow speeds because of the large temporal bandwidth associated with the exponential impulse response function as evident in Figure 4b. In principle it is possible to exploit the tools developed in above, such as multiple levels and denser connections, to remove the ringing. Unfortunately,

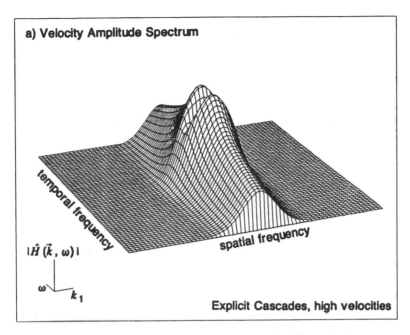

Figure 8a. The amplitude spectrum for a velocity filter selective to high speeds (flicker) constructed from CS filter using only two layers of local sums.

the use of multiple levels of simple sums causes a degradation in localization which imposes an unacceptable temporal delay, and an inordinate amount of added storage. Similarly, the arbitrary use of denser connections in explicit methods to obtain temporal smoothing requires too much storage. Therefore, in the case of low speed selective filters it is desirable to first, or simultaneously, apply exponential, low-pass, temporal smoothing. The exponential is convenient since it can be efficiently realized as a simple difference equation (see Eq. 28).

Local differences may also be used to enhance velocity specificity, while not severely degrading localization or increasing the computational expense and storage requirements. In particular, when designing filters selective to higher speeds, local differences in time can be used as a high-pass filter to remove sensitivity about the spatial frequency plane, $\omega = 0$,

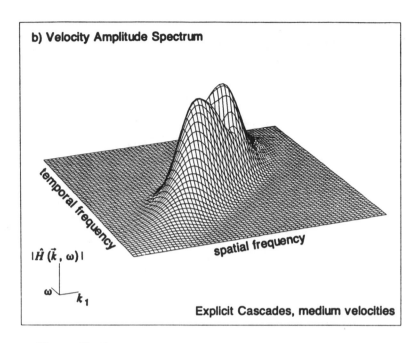

Figure 8b. The amplitude spectrum for a velocity filter selective to moderate speeds constructed from the CS filter using only two layers of local sums.

and therefore, for slow speeds and stationary patterns.

Examples of Orientation Selective Filters

To illustrate the filter properties more clearly we have constructed crude exemplary units similar to those discussed above. We show impulse responses, i.e., receptive fields, amplitude spectra, and the response behavior when applied to a simple synthetic stimulus. In the figures white corresponds to positive, grey to zero, and black to negative values. Details of the implementation are discussed in (Fleet and Jepson, 1985a). Also note that impulse responses and amplitude spectra have been scaled for visibility.

Figures 9a and 9b show the impulse responses for two orientation selective units constructed using two layers of simple sums in cascade with

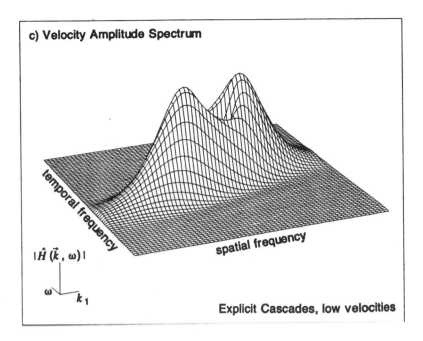

Figure 8c. The amplitude spectrum for a velocity filter selective to low speeds constructed from the CS filter using only two layers of local sums.

a DOG. Figures 9c and 9d show their amplitude spectra (cf. Figure 6a); half-sensitivity contours are shown in black. Adding a third layer of local sums in cascade with the filters shown in Figure 9 gives those shown in Figure 10 (cf. Figure 6b). Notice the increase in orientation specificity, the corresponding loss of localization, and the further attenuation of ringing in the amplitude spectra. To further demonstrate the tuning of these filters we have applied them to a simple stimulus consisting of a bright disc on a dark background. Figure 11 shows the stimulus and the response of a DOG. Figure 12 then shows the responses of the filters shown in Figures 9 and 10. To help appreciate the concentration of energy in the response patterns, i.e., the degree of orientation tuning, we have placed a black contour at half the maximum response (for positive responses only). The contours support the previous observation that the energy in the response of

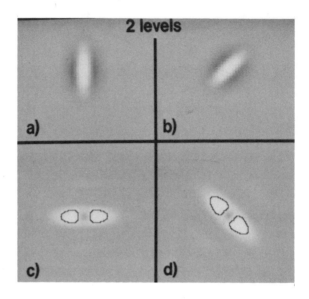

Figure 9. Impulse responses, a and b, and amplitude spectra, c and d, respectively, for orientation selective units constructed from the DOG with two layers of local sums.

the three level units is concentrated to a greater degree near a specific orientation.

Figure 13 shows the impulse responses and amplitude spectra resulting from the local difference operator, Eq. 33, being applied to the filters shown in Figure 9. Notice that: *i)* the differences do not seriously worsen localization since they only widen the impulse response in the direction perpendicular to the longest axis; and, *ii)* the differences change the phase by $\pi/2$. Figure 14 shows the responses of these filters to the disc stimulus.

In Figure 15 we show the impulse response and corresponding amplitude spectrum for an instance of the CS filter followed by a temporal exponential low-pass filter, Eqs. 22 and 23, thus giving a more isotropic spectrum. As with Figures 4b and 8 we only show one spatial dimension, the horizontal axis, and time, the vertical axis, in Figure 15. In Figures 16

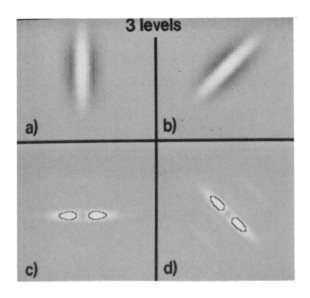

Figure 10. Impulse responses, a and b, and amplitude spectra, c and d, respectively, for orientation selective units constructed from the DOG with three layers of local sums.

and 17 the impulse responses and amplitude spectra for eight velocity selective filters are shown. They are tuned primarily to a) stationary components, b) slow speeds right, c) medium speeds right, d) fast speeds right, e) flicker, f) fast speeds left, g) medium speeds left, and h) slow speeds left. In the following demonstration and discussion it is important to note that the sensitivities of these eight filters overlap significantly. This is clear from Figure 18 in which we show the half-bandwidth contours superimposed on the band-pass amplitude spectrum of Figure 15b. Figure 18a shows the contours for filters a), c), e), and g). Figure 18b shows the contours for the other four, namely, b), d), f), and h).

To demonstrate the response behavior of these filters we applied them to the bright disc moving to the right at two pixels per frame across the dark background. In examining the responses note that different

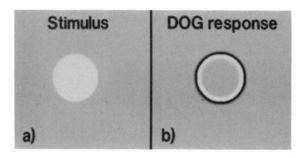

Figure 11. An example stimulus, a, of a bright disk on a dark background and response, b, of a radial band-pass DOG filter.

orientations move with difference speeds. Here, the speed of an oriented component depends on the cosine of the angle between it and the vertical component which is perpendicular to the direction of movement. Vertical components move at 2 pixels per frame. Components at $45°$ move at $\sqrt{2}$ pixels per frame, and horizontal components are stationary. In general, as given by Eq. 9b, a two-dimensional pattern translating with velocity, $\vec{v} = (v_1, v_2)$, can have non-zero frequency components only in a plane through the origin. Such a plane contains a one parameter family of normal velocities,

$$\vec{v}_n(\theta) = |\vec{v}\cos(\theta)| \, \vec{n}(\theta) , \qquad (40)$$

where $\vec{v}_n(\theta)$ has speed, $|\vec{v}\cos(\theta)|$, θ measures the angle between the direction of $\vec{v}$ and the normal direction, and the unit normal direction $\vec{n}(\theta)$ is given by

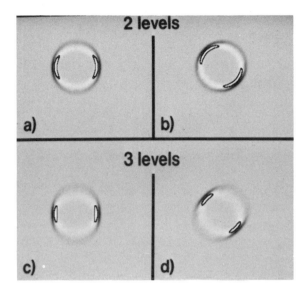

Figure 12. Orientation selective responses to the stimulus of Figure 11a with a and b corresponding to the units in Figures 9a and b, and c corresponding to Figures 10a and b.

$$\vec{n}(\theta) = \begin{bmatrix} \cos(\theta) & \sin(\theta) \\ -\sin(\theta) & \cos(\theta) \end{bmatrix} \begin{bmatrix} v_1 \\ v_2 \end{bmatrix} \frac{\text{sgn}(\cos(\theta))}{|\vec{v}|} . \tag{41}$$

Figure 19 shows the response of four velocity selective filters tuned to vertical orientations. These units were constructed using three layers of simple sums and one layer of differencing. Their speed preferences are for a) flicker, b) fast speeds right (about 2.5 pixels/frame), c) moderate speeds right (about 1 pixel/frame), and d) slow speeds right (about 0.4 pixels/frame). The other four channels, namely, those with preference to stationary patterns and leftward motion, give negligible responses. To appreciate the response differences among the filters we have plotted a contour at 50% of the maximum response of the channel with the greatest response. The contours were drawn only about the positive peaks to facilitate visibility. The negative peak on the right side of the disk is roughly as

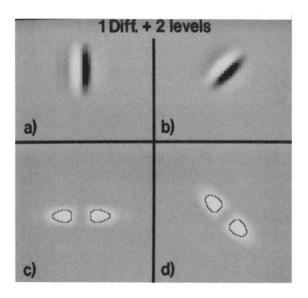

Figure 13. Impulse responses, a and b, and amplitude spectra, c and d, respectively, for units with local differences applied to the filters constructed with two layers of local sums as shown in Figures 9a and b.

strong as the positive peak on the left. Note that the filter tuned to fast speeds, Figure 19b dominates the response. There are no contours on the other three responses because as their maximum values fall short of half the maximum response of the fast speed channel shown in Figure 19b.

In Figure 20 we show the responses of filters selective to obliquely oriented patterns, with speed preferences, a) flicker, b) fast speeds right, c) moderate speeds right, and d) slow speeds right. According to Eq. 40 the speed of the oblique components is $\sqrt{2}$ pixels per frame which falls between the speeds preferred by the filters shown in Figures 20b and 20c Both of these channels and the next lower speed channel, Figure 20d, give substantial responses. It is also interesting to note that the filter selective to faster speeds, Figure 20b, is responding primarily to orientations less than 45°, while the filter selective to moderate speeds, Figure 20c, selects

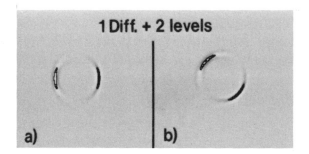

Figure 14. The responses of the difference operators shown in Figure 13 to the stimulus of Figure 11a. Note the removal of energy perpendicular to the preferred orientation as compared with Figures 12a and 12b.

orientations about 45° or more. This behavior is more pronounced in Figure 21 which shows the responses of four filters tuned to horizontal components with preferential tuning to a) slow speeds upward, b) stationary patterns, c) slow speeds downward, and d) moderate speeds downward. The filter tuned to stationary horizontal components shown in Figure 21b gives the largest response. The filters tuned to slow speeds up and down also repond, but select orientation components only to the right and the left of the horizontal, respectively.

Discussion

It essential to note that, apart from their conceptual simplicity and computational efficiency, the filters considered here have not been optimized. Given the variety of tools discussed in Sections 11.3.4 it is of

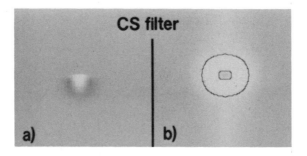

Figure 15. The impulse response, a, and amplitude spectrum, b, for the CS filter followed by a low-pass, exponential, temporal filter.

interest to examine which combinations of these tools are more near optimal in terms of the localization/specificity trade-off, while avoiding any appreciable ringing in the amplitude spectrum. It is also of interest to note that multiple levels may be coded implicitly through a very small number of connections where the convergence to orientation tuning occurs through iterations of lateral interactions within a level (Fleet and Jepson, 1985a). The use of implicit methods in time, however, suffer from the same problems as does the use of multiple levels, i.e., there is an unacceptably large temporal delay and an excessive amount of storage required. Thus, in addition, the relative efficiencies of implicit and explicit layers should be examined in terms of computational effort, the gain in orientation specificity, and the loss of spatial localization.

The approach demonstrated here of studying the signal structure produced by various image phenomena is currently being used to develop

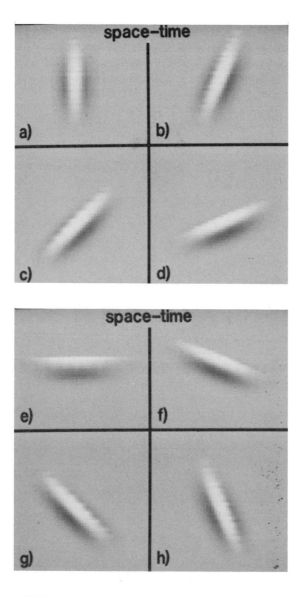

Figure 16. Impulse responses for eight velocity selective units with preferences to patterns exhibit the following types of movement: stationary, a; slow, moderate and fast speeds to the right, b, c and d; flicker, e; and fast, moderate and slow speeds to the left, f, g and h.

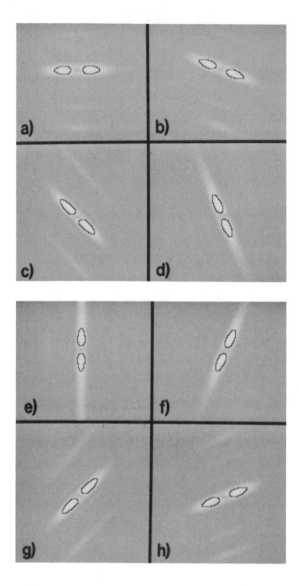

Figure 17. The amplitude spectra, respectively, for the impulse responses shown in Figure 16.

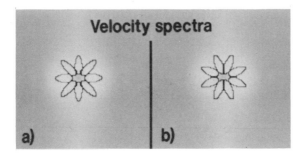

Figure 18. The half-sensitivity bandwidth contours from Figure 17 are superimposed on the band-pass spectrum shown in Figure 15b. The contours from Figures 17a, c, e and g are used in Figure 18a, while contours from Figures 17b, d, f, and h are used in Figure 18b.

blind processes for the extraction of other types of primitives. In particular, we are considering windowed linear filters for the measurement of information near occlusion boundaries (Jepson and Fleet, 1986). Nonlinear operators are being considered for the measurement of binocular disparity information, and similar operators could be considered for the measurement of local relative velocities.

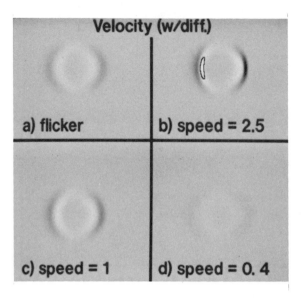

Figure 19. Responses to the stimulus of Figure 11a moving to the right at 2 pixels per frame by four units, all tuned to vertical orientations, with the following velocity preferences: flicker, a; fast to right, b; moderate to right, c; and slow to right, d. Note the dominance of the fast speed channel (Figure 19b).

11.4 Biological Research

The computational framework and filters developed in this chapter are not intended to model biological systems. We are, however, biologically motivated in several important respects, and it is of interest to point out those results that coincide with, or which might account for, physiological or psycho-physical observations.

It is encouraging to note that the simultaneous extraction of different modalities appears to be a feature of early visual processing in biological systems. In the human visual system the interpretation of spatial form often relies on the previous extraction of motion and depth information. This is exhibited in experiments with random dot patterns in which individual frames possess no apparent spatial coherence. When two frames are

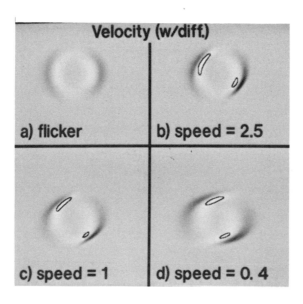

Figure 20. The responses of four units tuned to oblique orientations with speed preferences shown on the figure.

shown, one with a square, central region displaced, the result is a perception of depth when presented binocularly, or motion when presented in temporal succession despite the lack of spatial cues (Anstis, 1970; Julesz, 1971). Other compelling demonstrations have been provided by Adelson (1982). Thus, form interpretation does not necessarily precede the determination of motion or disparity. This is further supported by electrophysiology since cells at an early stage of processing are selectively sensitive to motion and disparity (Goodwin, et al., 1975; Movshon, et al., 1985; Poggio and Poggio, 1984).

The general style of hierarchical processing should be familiar to workers on biological vision. The use of bottom-up connections between layers and lateral interactions within layers is well established, although the present use of linearity is clearly an oversimplification. Regarding the extraction of orientation information we would be remiss not to point out

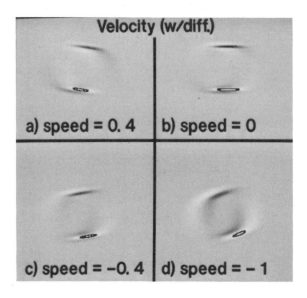

Figure 21. Responses of four units tuned to horizontal orientations
with several velocity preferences.

the similarity of the simple alignment of neighboring, isotropic, center-
surround units in Section 11.3.4 with the scheme proposed by Hubel and
Weisel (1962). It is important consider such a simple scheme with the
appropriate analysis, especially in the light of claims by Daugman (1983)
that such an approach cannot be successful.

   More recent research on orientation selectivity in the visual cortex has
been focused on the structure of receptive fields and frequency analysis.
The receptive fields of various simple cells have been modeled with two-
dimensional Gabor functions (Marcelja, 1980; Kulikowski, et al., 1982;
Pollen and Ronner, 1983). The general structure of such receptive fields is
elongated along the preferred orientation with a Gaussian-like envelope and
an aspect ratio of about 2. In the orientation orthogonal to the preferred
orientation, the receptive fields have a single excitatory central band and
inhibitory bands on both sides. In other cases, there are multiple stripes of

alternating excitatory and inhibitory inputs, sometimes symmetrical and other times antisymmetrical.

The simple alignment of isotropic band-pass units in Section 11.3.4 corresponds with examples of orientation cells but does not account for more than one excitatory and two inhibitory stripes or for antisymmetries. Interestingly, in Section 11.3.4 it was shown that local differences in the direction perpendicular to the original alignment are useful for the enhancement of orientation specificity. We also note that the use of such local differences introduces multiple sidebands, and each stage effectively alters the phase by $\pi/2$. This changes the structure of the receptive field from symmetrical to antisymmetrical, or *vice versa*, across the modulating profile. Similar results can be obtained through lateral interactions. Thus, by using local sums, local differences, and lateral interactions, we can easily obtain receptive field structures qualitatively similar to those currently modelled with Gabor functions.

Regarding the sensing of image motion we begin by noting that the spatio-temporal center-surround mechanism was developed as a first step towards modeling retinal processing. The CS model, while effectively an extension of the DOG model, includes explicit temporal properties that agree more closely with established physiology.

The properties of velocity selective mechanisms in the cortex are not as well established as the neural structure of orientation selective units, and are, therefore, more difficult to compare with the present filters. However, based in part on physiology and theoretical considerations, such mechanisms, expressed roughly in frequency space, have been suggested. Their use as interpolation filters to account for aspects of spatio-temporal hyperacuity has been suggested (Fahle and Poggio, 1981), while it has been shown that such filters provide a convenient account of short-range apparent motion phenomena (Morgan, 1980). More recently, models of human motion perception based on a frequency-space representation have been suggested (Adelson and Bergen, 1985; Watson and Ahumada, 1985), also see (van Santen and Sperling, 1985; Wilson, 1985). Of particular interest is physiological evidence suggesting separable channels in the cortex with fine spatial tuning and broad temporal tuning (Movshon, et al., 1985). Comparable results are reported in pycho-physics (Watson and Robson, 1981; Thompson, 1984). This would appear counter intuitive, and it would be of interest to examine how readily velocity might be extracted from such an encoding.

11.5 Machine Research

In accordance with the constraints on the first functional level of processing outlined in Section 11.3, current approaches to the extraction of motion information should be viewed as comprising several major levels of processing. Furthermore, they are typically separable into spatial and temporal stages of processing (Fleet and Jepson, 1985c). Usually, the first level performs a purely spatial analysis of the input, after which temporal properties are considered over a small number (usually 2) of frames. Perhaps the clearest examples are token-matching techniques (Ullman, 1979; Barnard and Thompson, 1980; Dreschler and Nagel, 1982) which consist of the two distinct stages: i) the extraction of tokens in isolated frames, and ii) the temporal matching, i.e., correspondence, of similar tokens in temporally adjacent frames. The first stage alone is a combination of several levels of processing which collectively perform a significant amount of spatial scene interpretation. These levels identify tokens, and remove noise and other irrelevant features. The second stage must infer a correspondence between similar tokens.

Similarly, gradient-based approaches (Limb and Murphy, 1975; Horn and Schunck, 1981; Nagel, 1983) involve multiple levels of processing, the first of which performs the extraction of spatial gradients. These gradients, along with temporal change information, are then used by later stages to compute normal velocities. We view the first stage of estimating spatial gradients as requiring interpretation since the solution to the motion constraint equation requires relatively significant and unique gradient values at a given image location. Velocity estimates are often restricted to specific spatial contours exhibiting special properties, such as high degrees of curvature, or consistency over several spatial scales. More importantly, the derivation of the motion constraint equation involves a restrictive assumption concerning the nature of the flow. In particular, in its common form, the constraint reflects the assumption of two-dimensional translational motion under parallel projection (Horn and Schunck, 1981). A different constraint equation has been derived in order to accomodate perspective projections and rotational motion (Schunck, 1984). However, this constraint relies on the assumption of "conservation of image features", which is not satisfied by natural motions of rough surfaces or of fluids. By the definitions outlined in Section 11.3.1 we consider these gradient-based approaches to be inferential.

In the present paradigm we propose the extraction of motion and depth information in the first functional level. Consequently, the first level should contain significant temporal and binocular components. In addition, in contrast to separable approaches, we emphasize that inseparable styles of processing are better suited to the extraction of orientation and normal velocity than are separable styles (Fleet and Jepson, 1985b). Our approach does not require an initial spatial analysis, in that it does not depend on the recognition of particular types of spatial structure, such as contours, peaks, or more elaborate features, in single frames of the raw image. We feel that this difference is extremely important, especially in domains where noise is a problem, since spatial features have proved to be difficult to obtain reliably. With velocity information measured directly in the initial stages of processing, it would appear that, in conjunction with spatial information, form interpretation should be simpler. Further desirable properties of spatio-temporal filtering are as follows. First, with frequency analysis we may evaluate the filters' orientation/velocity tuning, and also ensure reasonable signal properties. Unnecessary aliasing may be avoided, and noise is suppressed by reducing ringing in the amplitude spectrum and sensitivity about the fold-over rate, i.e., half the sampling rate. Secondly, these filters will respond acceptably to a wide variety of spatial patterns. Problematic areas for gradient-based approaches such as dense textures, are handled effectively. Semi-transparent surfaces and a background moving independently may be discerned since several velocity selective filters may respond at any spatio-temporal location.

Finally, it is essential to note that the filters presented here are not estimating optical flow. Rather, they provide estimates of the distribution of local motion information in a form which makes visual cues available for preliminary segmentation and should facilitate the determination of optical flow for a wide class of three-dimensional motions. We anticipate that these subsequent processes will involve a significant degree of interpretation. We also emphasize that a global Fourier transform or a power spectrum analysis is not being performed on the image, rather, frequency analysis is used only for the design of suitable filters.

References

Adelson, E.H., (1982) 'Some new illusions and some old ones, analyzed in terms of their Fourier components,' *Investigative Ophthalmology and Visual Science,* vol. 22, pp. 144.

Adelson, E.H., and Bergen, J.R., (1985) 'Spatiotemporal energy models for the perception of motion,' *J. Optical Society of America,* vol. 2, pp. 284-299.

Anstis, S.M., (1970) 'Phi movement as a subtractive process,' *Vision Research,* vol. 10, pp. 1411-1430.

Ballard, D., (1985) 'Cortical connections and parallel processing: Structure and function,' *The Behavioral and Brain Sciences,* pp. 1-43.

Barnard, S.T. and Thompson, W.B., (1980) 'Disparity analysis of images,' *IEEE Trans. on Pattern Analysis and Machine Intelligence,* vol. PAMI-2, pp. 333-340.

Barrow, H., and Tenenbaum, J.M., (1978) 'Recovering intrinsic scene characteristics from images,' in **Computer Vision Systems,** A. Hanson and E. Riseman (eds.), Academic Press, New York.

Brillouin, L., (1962) **Science and Information Theory,** 2nd Edition, Academic Press, New York.

Burt, P.J., (1981) 'Fast filter transforms for image processing,' *Computer Graphics Image Processing,* vol. 16, pp. 20-51.

Capek, M., (1981) 'Time in relativity theory: Arguments for a theory of becoming,' in **The Voices of Time,** J.T. Fraser (ed.), The University of Massachusetts Press, Amherst, MA.

Cowey, A., (1979) 'Cortical maps and visual perception,' *Quart. J. of Experimental Psychology,* vol. 31, pp. 1-17.

Crowley, J.L., (1982) 'Representation for Visual Information,' Carnegie Mellon University, PhD. Dissertation. Available as CMU Robotics Institute Tec. Rep. CMU-RI-TR-82-7.

Daugman, J.D., (1983) 'Six formal properties of two-dimensional anisotropic visual filters: Structural principles and frequency/orientation selectivity,' *IEEE Trans. Systems, Man and Cybernetics,* vol. SMC 13, pp. 882-887.

Dreschler, L.H., and Nagel, H.-H., (1982) 'Volumetric model and 3D trajectory of a moving car derived from monocular TV frames of a street scene,' *Computer Graphics Image Processing,* vol. 20, pp. 199-228.

Estes, W., (1978) 'Perceptual processing in letter recognition and reading,' in **The Handbook of Perception: Perceptual Processing,** vol. IX, E. Carterette and M. Friedman (eds.), Academic Press, New York.

Fahle, M., and Poggio, T., (1981) 'Visual hyperacuity: Spatio-temporal interpolation in human vision,' *Proc. Royal Society London,* vol. B 213, pp. 451-477.

Fleet, D.J., Hallett, P.E., and Jepson, A.D., (1985) 'Spatio-temporal inseparability in early visual processing,' *Biological Cybernetics,* vol. 52, pp. 153-164.

Fleet, D.J., and Jepson, A.D., (1984) 'A cascaded filter approach to the construction of velocity selective mechanisms,' Tech. Rep. RBCV-TR-84-6, Dept. of Computer Science, University of Toronto.

Fleet, D.J., and Jepson, A.D., (1985a) 'On the hierarchical construction of orientation and velocity selective filters,' Tech. Rep. RBCV-TR-85-8,

Dept. of Computer Science, University of Toronto.

Fleet, D.J., and Jepson, A.D., (1985b) 'Spatiotemporal inseparability in early vision: Center-surround models and velocity selectivity,' *Computational Intelligence,* vol. 1, pp. 89-102.

Fleet, D.J., and Jepson, A.D., (1985c) 'Velocity extraction without form interpretation,' *Proc. 3rd IEEE Workshop on Computer Vision: Representation and Control,* Bellaire, MI, pp. 179-185.

Ganz, L., (1975) 'Temporal factors in visual perception,' in **The Handbook of Perception: Seeing,** vol. V, E.C. Carterette and M.P. Friedman (eds.), Academic Press, New York.

Goodwin, A.W., Henry, G.H., and Bishop, P., (1975) 'Direction selectivity of simple striate cells: Properties and mechanisms,' *J. Neurophysiology,* vol. 38, pp. 1500-1523.

Gross, C.G., (1973) 'Visual functions of inferotemporal cortex,' in **Handbook of Sensory Physiology,** vol. VII, 3B, R. Jung (ed.), Springer-Verlag, New York, pp. 451-482.

Horn, B.K.P, and Schunck, B.G., (1981) 'Determining optic flow,' *Artificial Intelligence,* vol. 17, pp. 185-204.

Hubel, D.H., and Weisel, T.N., (1962) 'Receptive fields, binocular interaction, and functional architecture in the cat's visual cortex,' *J. Physiology,* vol. 160, pp. 106-154.

Jepson, A.D. and Fleet, D.J., (1987) 'The measurement of visual primitives,' in preparation.

Julesz, B., (1971) **Foundations of Cyclopean Perception,** University of Chicago Press, Chicago.

Kiroussis, L., and Papadimitriou, C., (1985) 'On the intractability of polyhedral labelling algorithms,' *Proc. Foundations of Computer Science.*

Kulikowski, J.J., Marcelja, S., and Bishop, P.O., (1982) 'Theory of spatial properties and spatial frequency relations in the receptive fields of simple cells in the visual cortex,' *Biological Cybernetics,* vol. 43, pp. 187-198.

Limb, J.O., and Murphy, J.A., (1975) 'Estimating velocity of moving images in television signals,' *Computer Graphics Image Processing,* vol. 4, pp. 311-327.

Marcelja, S., (1980) 'Mathematical descriptions of responses of simple cells,' *J. Optical Society America,* vol. 70, pp. 1297-1300.

Marr, D., and Hildreth, E.C., (1980) 'Theory of edge detection,' *Proc. Royal Society London,* vol. B 207, pp. 187-217.

Marr, D., and Ullman, S., (1981) 'Directional selectivity and its use in early visual processing,' *Proc. Royal Society London,* vol. B 211, pp. 151-180.

McKee, S.P., (1981) 'A Local Mechanism for Differential Velocity Detection,' *Vision Research,* vol. 21, pp. 491-500.

Morgan, M.J., (1980) 'Analogue models of motion perception,' *Phil. Trans. Royal Society London,* vol. B 290, pp. 117-135.

Movshon, J.A., Adelson, E.H., Gizzi, M.S., and Newsome, W.T., (1985) 'The analysis of moving visual patterns,' in **Pattern Recognition Mechanisms**, C. Chages, R. Gattass and C. Gross (eds.), Vatican Press, Rome, pp. 117-151.

Nagel, H.-H., (1983) 'Displacement vectors derived from second-order intensity variations in image sequences,' *Computer Graphics and Image Processing,* vol. 21, pp. 85-117.

Nakayama, K., (1985) 'Biological image motion processing: A review,' *Vision Research,* vol. 25, pp. 625-659.

Neisser, U., (1967) **Cognitive Psychology,** Appleton-Century-Crofts, New York.

Poggio, G., and Poggio, T., (1984) 'The analysis of stereopsis,' *Annual Rev. Neuroscience,* vol. 7, pp. 379-412.

Pollen, D.A., and Ronner, S.F., (1983) 'Visual cortical neurons and localized spatial frequency filters,' *IEEE Trans. on Systems, Man and Cybernetics,* vol. SMC-13, pp. 907-916.

Potter, M.C., (1975) 'Meaning in visual search,' *Science,* vol. 187, pp. 965-966.

Rabiner, L.R., and Gold, B., (1975) **Digital Signal Processing.** Prentice-Hall, Englewood Cliffs, NJ.

Richter, J., and Ullman, S., (1982) 'A model for the temporal organization of X and Y type receptive fields in the primate retina,' *Biological Cybernetics,* vol. 43, pp. 127-145.

Scholl, D., (1956) **The Organization of the Cerebral Cortex,** Wiley.

Schunck, B.G., (1985) 'Image flow: Fundamentals and future research,' *IEEE Conf. on Computer Vision and Pattern Recognition,* San Francisco, pp. 561-571.

Slepian, D., (1983) 'Some comments on Fourier analysis, uncertainty and modelling,' *SIAM Rev.*, vol. 25, pp. 379-393.

Sperling, G., (1963) 'A model for visual memory tasks,' *Human Factors,* vol. 5, pp. 19-31.

Stone, J., Dreher, B., and Leventhal, A., (1979) 'Hierarchical and parallel mechanisms in the organization of the visual cortex,' *Brain Research Reviews,* vol. 1, pp. 345-394.

Thompson, P., (1984) 'The coding of the velocity of movement in the human visual system,' *Vision Research,* vol. 24, pp. 41-45.

Treisman, A.M., and Schmidt, H., (1982) 'Illusory conjunctions in the perception of objects,' *Cognitive Psychology,* vol. 14, pp. 107-141.

Tsotsos, J.K., (1988) 'How does human vision beat the time complexity of visual perception?,' to appear in **Computational Processes in Human Vision: An Interdiciplinary Approach**, Z. Pylyshyn (ed.), Ablex Press.

Ullman, S., (1979) **The Interpretation of Visual Motion**, MIT Press, Cambridge, MA.

Uttal, W., (1981) **A Taxonomy of Visual Processes**, Lawrence Erlbaum.

van Essen, D., and Maunsell, J., (1983) 'Hierarchical organization and functional streams in the visual cortex,' *Trends in Neuroscience,* pp. 370-375.

van Santen, J.P.H., and Sperling, G., (1985) 'Elaborated Reichardt detectors,' *J. Optical Society America,* vol. A 2, pp. 300-321.

Watson, B.A., and Ahumada, A.J., Jr., (1985) 'Model of human visual-motion sensing,' *J. Optical Society America,* vol. A 2, pp. 322-342.

Watson, B.A., and Robson, J.G., (1981) 'Discrimination at threshold: Labelled detectors in human vision,' *Vision Research*, vol. 21, pp. 1115-1122.

Watson, A., Thompson, P., Murphy, B., Nachmais, J., (1980) 'Summation and discrimination of gratings moving in opposite directions,' *Vision Research*, vol. 20, pp. 341-347.

Wilson, H.R., (1985) 'A model for directional selectivity in threshold motion perception,' *Biological Cybernetics*, vol. 51, pp. 213-222.

AUTHOR INDEX

SUBJECT INDEX